GROUNDWATER PUMPING TESTS

Design & Analysis

WILLIAM C. WALTON

Library of Congress Cataloging-in-Publication Data

Walton, William Clarence.
Groundwater pumping tests.

Bibliography: p.
Includes index.
1. Groundwater flow—Measurement. 2. Aquifers—Measurement. I. Title. II. Title: Groundwater pumping tests.
GB1197.7.W34 1987 551.49 87-2950
ISBN 0-87371-108-4

Second Printing 1988

LEWIS PUBLISHERS, INC.
121 South Main Street, Chelsea, Michigan 48118

PRINTED IN THE UNITED STATES OF AMERICA

Preface

Practical aspects of, guidelines, and techniques for groundwater pumping test design and analysis are covered in this book. Students and professionals who are concerned with the proper conduct and interpretation of a pumping test are target readers. This book was written specifically for practicing engineers, geologists, and hydrogeologists; water well contractors; educators and students; governmental personnel; industry representatives; and others. The reader is expected to have a working knowledge of hydrogeology or access to books on groundwater geology and hydrology.

Too frequently, groundwater pumping test design and analysis ignore well storage capacity, delayed gravity yield, well partial penetration, and aquitard storativity impacts without proving them negligible. As a result, erroneous conclusions are reached concerning aquifer system hydraulic characteristics, boundaries, and discontinuities. Pumping test data often is filtered arbitrarily without adequate justification in attempts to match inappropriate aquifer models and field conditions. Messages sent by the aquifer system through water level data sometimes are misinterpreted or missed. Antecedent water level trends and water level adjustments for changes in barometric pressure and surface water stages frequently are ignored in calculating drawdown and recovery. Finally, manual graphic analysis supplemented with microcomputer programs is, to an excessive extent, being replaced by fully automatic microcomputer analysis without critical examination of interpretative methods in program algorithms and their limitations. Hopefully, this book will focus needed attention on the facets mentioned above and improve the accuracy and reliability of pumping test design and analysis.

This book grew from my lecture notes for a Seminar On Techniques For Pumping Test Design And Analysis presented on February 10–11, 1986, at the Hilton University Inn, Columbus, Ohio. This Seminar was part of the National Water Well Association Distinguished Seminar Series on Ground Water Science For 1986. Comments and reactions of the 50 people participating in the Seminar were most helpful in the preparation of this book.

William C. Walton

Mahomet, Illinois

William C. Walton is a semi-retired Consultant In Water Resources and resides on the banks of the Sangamon River at Mahomet, Illinois. Bill received his B.S. in Civil Engineering from Lawrence Institute of Technology in 1948 and attended Indiana University, University of Wisconsin, Ohio State University, and Boise State University. He served for 10 years as Director of the Water Resources Research Center and Professor of Geology and Geophysics at the University of Minnesota.

Bill has 39 years of experience in the water resources field including 1 year with the U. S. Bureau of Reclamation, 7 years with the U. S. Geological Survey, 6 years with the Illinois State Water Survey, and 9 years with consulting companies such as Camp Dresser & McKee, Inc. and Geraghty & Miller, Inc. He was Executive Director of the Upper Mississippi River Basin Commission for 5 years. Bill has participated in water resources projects throughout the United States and Canada and in Haiti, El Salvador, Libya, and Saudi Arabia.

He is an honorary Life Member and past Vice President of the National Water Well Association and a past editor of the Journal of Ground Water. Bill is past Chairman of the Ground Water Committee of the Hydraulics Division, American Society of Civil Engineers and a past member of the U. S. Geological Survey Advisory Committee on Water Data for Public Use. He is a past Consultant to the Office of Science and Technology, Washington, D.C. and an Advisor to the United States Delegation to the Coordinating Council of the International Hydrological Decade of UNESCO.

Bill is the author of three books and over 70 technical papers. He served as a Visiting Scientist for the American Geophysical Union and the American Geological Institute and lectured at many universities throughout the United States.

Contents

List of Tables

List of Figures

List of Microcomputer Programs

1

Introduction

During the last two decades, great progress has been made in computer modeling of groundwater flow and contaminant transport problems. The reliability of model results largely rests on knowledge of the hydraulic characteristics of aquifer systems. A pumping test is one of the most suitable means of quantifying hydraulic characteristics because it yields results which, in general, are representative of a larger area than are results from single point observations.

A pumping test is herein defined as a field in situ study aimed at obtaining controlled aquifer system response data. Usually, a production well is pumped at several fractions of full capacity and/or at a constant rate, and water levels are measured at frequent intervals in the production well and nearby observation wells. Time-drawdown and distance-drawdown data are analyzed with model equations and type-curve matching, straight-line matching, or inflection-point selection techniques.

In this book, pumping test techniques presented by Ferris et al. (1962); Walton (1962); Bentall (1963); Walton (1970); Stallman (1971); Lohman (1972); Neuman (1974); Kruseman and De Ridder (1976); and Reed (1980) are modified, supplemented, and integrated with microcomputer programs to facilitate improved pumping test design and analysis. Type-curve matching techniques are emphasized. Methods for analyzing the response of an aquifer to an instantaneous charge of water (Cooper et al., 1967, pp.

263–269) and the head variation due to depressurization (Bredehoeft and Papadopulos, 1980, pp. 233–238) are not covered. This book restricts itself to pumping tests in uniformly porous aquifer systems such as granular deposits. The reader is referred to Streltsova-Adams (1978, pp. 357–423) for information on well hydraulics in fractured rock systems.

Emphasis has been placed in this book on practice rather than theory. Pumping test analysis equations are presented in abbreviated and final form using the gallon-day-foot system of units. Abbreviations used are: feet (ft), inches (in.), seconds (sec), minutes (min), gallons per minute (gpm), cubic feet per second (cfs), gallons per day per foot (gpd/ft), gallons per day per square foot (gpd/sq ft), gallons per minute per foot (gpm/ft), and gallons per day per acre per foot (gpd/acre/ft). Mathematical derivations of equations may be pursued through references. Application of techniques to problem solving is stressed. Many example problems are included, with step-by-step solutions. This book is intended as a self-contained guidebook aimed at the practical worker; subject matter is presented in as non-mathematical a manner as possible.

A variety of aquifer system and facility conditions may influence pumping test data, including:

1. production well discharge partly derived from water stored within the well (well storage capacity)
2. partially penetrating wells with aquifer stratification
3. water table decline with delayed gravity yield

Erroneous conclusions may be reached if these conditions are ignored in pumping test analysis. Condition impacts on time-drawdown and distance-drawdown data are site and facility dependent. They may be either negligible or appreciable, as indicated by the following equations (see Driscoll, 1986, p. 566; Hantush, 1964, p. 351; and Walton, 1962, p. 6):

$$t_s = 5.4 \times 10^5 (r_w^2 - r_c^2)/T \quad (1.1)$$

$$r_p = 1.5m (P_H/P_V)^{1/2} \quad (1.2)$$

$$t_d = 5.4 \times 10^4 mS_y/P_V \quad (1.3)$$

where t_s = time after pumping started beyond which well storage capacity impacts are negligible (less than 1% of drawdown values), in min

r_w = production well effective radius, in ft

r_c = pump-column pipe radius, in ft

T = aquifer transmissivity, in gpd/ft

r_p = distance from production well beyond which partial penetration impacts are negligible, in ft

m = aquifer thickness, in ft

P_H = aquifer horizontal hydraulic conductivity, in gpd/sq ft

P_V = aquifer vertical hydraulic conductivity, in gpd/sq ft

t_d = time after pumping started beyond which delayed gravity yield impacts are negligible, in min

S_y = aquifer water table storativity (specific yield), dimensionless

(The microcomputer programs included with this book will solve Equations 1.1–1.3, as well as all other equations cited in the book, based on input provided by the user. Printouts of the programs are provided in Appendix A.) Examples of impact importance for three common site and facility conditions are presented in Table 1.1.

In general, well storage capacity impacts tend to be negligible, except for the first few minutes of the pumping period, with moderate to high (>10,000 gpd/ft) transmissivities and small (<1 ft) well radii. Appreciable well storage capacity impacts occur with low transmissivities and/or large well radii. Partially penetrating well impact distances increase with increases in aquifer thickness and the ratio P_H/P_V from less than 10 ft to more than 100 ft

Table 1.1. Impacts of Site and Facility Conditions

$T = 100{,}000$ gpd/ft	$T = 10{,}000$ gpd/ft	$T = 1000$ gpd/ft
$S_y = 0.20$	$S_y = 0.10$	$S_y = 0.05$
$P_H = 5000$ gpd/sq ft	$P_H = 200$ gpd/sq ft	$P_H = 10$ gpd/sq ft
$P_V = 1000$ gpd/sq ft	$P_V = 40$ gpd/sq ft	$P_V = 5$ gpd/sq ft
$m = 20$ ft	$m = 50$ ft	$m = 100$ ft
$r_w = 0.50$ ft	$r_w = 0.33$ ft	$r_w = 2$ ft
$r_c = 0.20$ ft	$r_c = 0.15$ ft	$r_c = 0.10$ ft
$t_s = 1.1$ min	$t_s = 4.7$ min	$t_s = 2.2 \times 10^3$ min
$r_p = 67$ ft	$r_p = 168$ ft	$r_p = 212$ ft
$t_d = 216$ min	$t_d = 6750$ min	$t_d = 5.4 \times 10^4$ min

from the production well. Time durations when delayed gravity yield impacts are appreciable increase with increases in aquifer thickness and decreases in aquifer vertical hydraulic conductivity from less than 100 minutes to more than 7 days. Particular attention should be paid to the conditions mentioned above.

Other important aquifer system and facility conditions which may influence pumping test data and are covered in this book include:

1. nearby aquifer system boundaries or discontinuities
2. aquitard storativity and source bed drawdown
3. decreased transmissivity with water table decline
4. interference from nearby production wells
5. changes in barometric pressure or stream stages
6. tidal fluctuations

Pumping test design should be guided by a pretest conceptual modeling effort which involves estimating hydraulic characteristics and boundary conditions with available hydrogeologic data and predicting aquifer system response to pumping. Optimal depths, locations, and number of observation wells, and pumping rate and duration are selected based on predicted aquifer system behavior. A microcomputer program is provided for use with pretest conceptual models; it is a modification of a numerical pumping test program developed in polar form by Rushton and Redshaw (1979, pp. 231–293). The program covers nonleaky artesian, leaky artesian, and water table conditions; well storage capacity; storativity conversion; delayed gravity yield; and decrease in transmissivity with water table decline. Another microcomputer program, based on equations given by Hantush (1964, p. 353), is provided for calculation of well partial penetration impacts in production and observation wells.

In pretest conceptual models, boundary and aquifer discontinuity impacts are simulated with image wells. Interference from nearby production wells is modeled with multiple-well system concepts. Whether or not boundary, discontinuity, or interference impacts are appreciable may be ascertained with the following equation (see Ferris et al., 1962, p. 93; set u = 5):

$$r_n = 4.3 \times 10^{-2}(Tt/S_{aw})^{1/2} \qquad (1.4)$$

where r_n = distance from observation well beyond which boundary and discontinuity image well or nearby production well impacts are negligible, in ft

S_{aw} = artesian or water table storativity (use 0.005 for leaky artesian), dimensionless

T = aquifer transmissivity, in gpd/ft

t = time after pumping started, in min

In general, boundary, discontinuity, or interference impacts with a pumping period of 1 day tend to be negligible when image or production wells are beyond distances of 1000, 5000, and 10,000 ft under water table, leaky artesian, and nonleaky artesian conditions, respectively. Under leaky artesian conditions, it is often necessary to collect and analyze data for aquitard wells. Equations and a table given by Witherspoon and Neuman (1972, p. 267) are utilized to calculate the vertical hydraulic conductivity of an aquitard by comparing drawdowns in aquifer and aquitard observation wells nested at the same site.

Drawdown and recovery should be estimated from pumping test data by extrapolating antecedent water level trends. Water level adjustments should be made for any barometric pressure, stream stage, and tidal fluctuations. Adjusted water level data may be matched to the well function type curve for the analytical model which best suits the specific aquifer system and facility conditions encountered in the field. Match-point coordinates may be inserted in model equations to quantify hydraulic characteristics (see Ferris et al., 1962).

The pretest conceptual model serves as a frame of reference for selecting the appropriate type curve and guides corroboration of the reliability of pumping test results. Microcomputer programs (see Walton, 1985, pp. 341–343) for estimating partial penetration and other selected well functions are provided to facilitate the construction of type curves for complex aquifer system and well conditions. Several pumping test case studies are presented to illustrate the application of techniques.

Pumping test analysis tends to be nonunique because field measurements are usually limited in accuracy, analytical models seldom completely simulate reality, and several combinations of hydraulic characteristic and boundary conditions may satisfy model equations. Analysis accuracies of 15% for hydraulic conductivity and 30% for storativity are commonly acceptable.

The microcomputer programs are written in IBM PC (trademark of International Business Machines Corp.) BASIC. They may also be run on Microsoft (trademark of Microsoft Corp.) BASIC on IBM compatible computers, or converted for use on non-IBM compatible computers (see Harris and Scofield, 1983). Source codes for the programs are provided in Appendix A. A diskette called GWPT and containing a complete set of the BASIC programs is included with this book. Complete instructions on how to run GWPT are presented in Appendix B. The diskette runs on personal computers such as the IBM PC and compatible machines that use the MS-DOS (trademark of Microsoft Corp.) operating system.

Appendix C contains tables with representative horizontal hydraulic conductivity, aquitard vertical hydraulic conductivity, aquifer vertical-to-horizontal hydraulic conductivity ratio, induced streambed infiltration rate, specific yield, and artesian storativity ranges for selected rocks to facilitate pretest conceptual modeling. A table of conversion coefficients for a number of units regularly encountered in pumping test design and analysis is provided in Appendix D. Values of selected type curve functions and water dynamic viscosity are extensively tabulated in Appendix E. A microcomputer program (PT3, Appendix A), based on the Lagrange equation and a program presented by Poole et al. (1981, pp. 84–85), is provided for interpolation of type curve functions between values given in tables.

2

Design and Field Observation

The purpose of pumping test design is to ensure that a proposed test site and associated equipment will yield acceptable results, and to minimize uncertainties in data collection and analysis. Criteria used in site evaluation are (Stallman, 1971, pp. 6–8):

1. The production well should be equipped with reliable power, pump, and discharge-control equipment.
2. The water discharged should be conducted away from the production well to minimize or eliminate recirculation.
3. The wellhead and discharge lines should be accessible for installing, regulating, and monitoring equipment.
4. It should be possible to measure water levels in the production well before, during, and after pumping.
5. The diameter, depth, and position of all intervals open to the aquifer system should be known, as should total depth.
6. All production wells within the pumping test area of influence should be capable of being controlled, and their discharges should be known.
7. Fluctuations in nearby surface water stages should be monitored.
8. Existing production and observation wells should be utilized whenever possible.
9. Response of observation wells to changing water level stages should be tested by injecting a known volume of water into each well and measuring the subsequent decline of water levels. The initial rise of water levels

should usually be dissipated within a few minutes if the observation well is to accurately reflect changes of head in the aquifer system during the test.

10. Radial distance and direction from the production well to each observation well and from any interfering production wells should be determined.
11. Radial distance and direction from any known boundaries to each observation well should be determined.
12. Depth to, thickness of, and areal and vertical limits of the aquifer system should be determined.
13. Nearby aquifer system discontinuities should be mapped.
14. The order of magnitude of pertinent aquifer system hydraulic characteristics should be estimated.

With induced streambed infiltration conditions, measurements of streambed width and depth of water in the stream should be made at frequent intervals at least 1000 ft up- and downstream from the production well.

GENERALIZED DESIGN FEATURES

Pumping tests involve large investments in time and money and, to some degree, design is dictated by nonscientific considerations. In general, at least three observation wells at various distances from the production well should be available. Observation well spacing should be logarithmic and designed to provide at least one logarithmic cycle of distance-drawdown data. A typical spacing is 100, 400, and 1000 ft. To avoid any partial penetration impacts, the distance from the closest observation well to the production well should be equal to or greater than

$$1.5m(P_H/P_V)^{1/2} \tag{2.1}$$

where m = aquifer thickness, in ft
P_H = aquifer horizontal hydraulic conductivity, in

gpd/sq ft
P_V = aquifer vertical hydraulic conductivity, in gpd/sq ft

If measurement of partial penetration impacts is important, an observation well should be closer to the production well at a distance equal to or less than the aquifer thickness.

With boundary conditions, observation wells should be spaced along a line through the production well and parallel to the boundary to minimize the effects of the boundary on distance-drawdown data. It is desirable to space observation wells also on a line perpendicular to the boundary and at variable distances and directions from the image well associated with the boundary. Cone of depression deviations from symmetry may be ascertained by locating observation wells on two radial lines at a right angle with one another.

Under induced streambed infiltration conditions, the production well should be located a distance equal to the aquifer thickness from the stream bank to minimize streambed partial penetration impacts. At least two observation wells should be located one logarithmic cycle apart on a line through the production well and parallel to the streambed. Distances from the production well to these two observation wells should be 50 and 500 ft, respectively. Two additional observation wells should be located on a line through the production well and at a right angle to the streambed. One observation well should be located at the near stream bank. The other observation well should be located across the stream and at a distance equal to the aquifer thickness from the far stream bank. (If the streambed width exceeds 500 ft, delete this observation well.)

The open (screened) portions of both production and observation wells are usually spaced vertically opposite the same aquifer zones. Fully penetrating wells are commonly desirable. With partial penetration conditions, at least one nest of observation wells is desirable near the production well. One observation well of the nest should be open opposite the aquifer zone open to the production well, and the other observation well of the nest should be open near the aquifer top or base. With leaky artesian conditions, at least one nest of aquifer and aquitard observation wells is desirable. The aquifer observation well should be open opposite the aquifer zone open to the production well, and the aquitard observation well should be open opposite a lower portion of the aquitard. Aquifer and aquitard observation well diameters usually exceed 1 in. and are often 4–6 in. when float-operated recorders are required.

An understanding and appreciation of the extent of the aquifer system volume affected by pumping facilitates pumping test design. Without boundaries, that volume is a cylinder with a radius equal to the distance from the production well beyond which drawdown is negligible (see Equation 1.4). Under nonleaky artesian and water table conditions, the aquifer system volume cylinder height affected by pumping is the aquifer saturated thickness. Under leaky artesian conditions, the cylinder height is the aquifer thickness plus overlying and underlying portions of aquitards and source beds within the vertical extent of the cone of depression. The vertical hydraulic conductivity of any window through an aquitard is integrated with the vertical hydraulic conductivity of the aquitard outside any window within the extent of the cone of depression.

Hydraulic characteristic values estimated with pumping test data are means of hydraulic characteristic values at many sites within a cylindrical volume. A point specific hydraulic characteristic value based on a well sample may deviate significantly from pumping test results. Correlation between pumping test and well sample hydraulic characteristic values should proceed with due caution.

Under water table conditions, the storativity and vertical hydraulic conductivity values calculated with pumping test data are mean values for the portion of the aquifer system volume between the initial water table and the cone of depression, not the entire aquifer thickness. The aquitard vertical hydraulic conductivity value calculated with pumping test data is the mean value for the portion of the aquitard affected by pumping, not necessarily the entire aquitard thickness. With partially penetrating wells, the aquifer vertical hydraulic conductivity value calculated with pumping test data is the mean value for the portion of the aquifer system volume having a radius equal to $1.5m(P_H/P_V)^{1/2}$.

Pumping tests are commonly 8 hours to 5 days in duration. With a tight aquitard, a pumping test without aquitard observation wells may have a duration of several months. If a boundary occurs within the aquifer system volume likely to be affected by pumping, the duration of the pumping test should be long enough so that drawdown deviations due to the boundary are clear. The appropriate duration of a test under artesian conditions may be determined with the following equation (rearrange Equation 1.4 and add one time logarithmic cycle):

$$t_i = 5.4 \times 10^3(r_i^2)S_{aw}/T \qquad (2.2)$$

where t_i = pumping test duration which must be exceeded if boundary impacts are to be clear (one time logarithmic cycle after impacts become appreciable), in min

r_i = distance from observation well to boundary image well, in ft

S_{aw} = aquifer artesian or water table storativity, dimensionless

T = aquifer transmissivity, in gpd/ft

The appropriate duration of a test under water table conditions may be determined from Equation 1.3 or 2.2, whichever gives the greater duration.

Time intervals for observation well water level measurements vary from short at the start of the test, when water level declines are rapid, to long at the end of the test, when the time rate of drawdown is small. A typical range of time

Table 2.1. Time Intervals for Observation Well Measurements

Time After Pumping Started	Time Intervals
1 - 2 minutes	10 seconds
2 - 5 minutes	30 seconds
5 - 15 minutes	1 minute
15 - 50 minutes	5 minutes
50 - 100 minutes	10 minutes
100 - 500 minutes	30 minutes
500 - 1000 minutes	1 hour
1000 - 5000 minutes	4 hours
5000 - end	1 day

intervals for observation well water level measurements is shown in Table 2.1.

A typical pumping test schedule for an artesian aquifer system is as follows:

Day 1. water level measurements to establish antecedent trend

Day 2. 1-hour trial test to adjust equipment followed by a 1-hour recovery period; 3-hour step-drawdown test to determine production well well loss coefficient followed by a 20-hour recovery period

Day 3. 24-hour constant rate test to determine aquifer system hydraulic characteristics and boundary conditions

Day 4. 24-hour recovery test to verify aquifer system hydraulic characteristics and boundary conditions

DETAILED DESIGN WITH PRETEST CONCEPTUAL MODEL

Detailed pumping test design is facilitated by a successive approximation pretest conceptual modeling effort which predicts the outcome of the test beforehand. In developing a pretest conceptual model, initial aquifer system hydraulic characteristics, boundaries, and discontinuities and well construction features are quantified based on available hydrogeologic data. The tables in Appendix C on representative hydraulic characteristics and text on general design features are useful in the modeling effort. Drawdown distribution with initial conceptual hydraulic characteristic values and boundary conditions is estimated with microcomputer programs. The initial pretest conceptual model is modified if selected well construction features and the duration of the test do not seem reasonable in light of the calculated drawdown distribution. The modified pretest conceptual model is used to recalculate drawdown distribution. The process is repeated until satisfactory results are obtained. Final design features are then selected in light of the pretest conceptual model results. A range of pretest conceptual models may be selected for analysis to bracket the range of possible aquifer response.

Microcomputer Programs

A conversion to IBM BASIC and a modification of a microcomputer program, written in TRS-80 (trademark of Tandy Corp.) BASIC by Rathod and Rushton (1984, pp. 602–608) specifically for pumping test analysis, may be utilized to estimate pretest conceptual model drawdown distribution. Modifications made by the author include expansions to cover leaky artesian and water table conditions (see Rushton and Redshaw, 1979; pp. 249–262) and incorporation of the relationship between the delay index and aquifer vertical hydraulic conductivity as discussed by Neuman (1975, p. 336). The modified microcomputer program (PT1) presented in Appendix A simulates radial two-dimensional flow toward a production well through a unit width aquifer system slice extending from the well to an outer boundary. The following aquifer system and well conditions are covered in the program: nonleaky artesian, leaky artesian, water table, delayed gravity yield, decreased transmissivity with water table decline, and storativity conversion. At any radius, hydraulic characteristics have the same value as they do at any depth. Although pump-column pipe impacts, aquitard storativity, and source bed drawdown are assumed to be negligible, calculated drawdowns are sufficiently accurate for pumping test design purposes.

The interactive database input section of the program prompts the user to enter all or some of the following aquifer system characteristics and well conditions, depending upon whether nonleaky, leaky artesian, or water table conditions prevail.

1. aquifer horizontal hydraulic conductivity, in gpd/sq ft
2. aquifer vertical hydraulic conductivity, in gpd/sq ft
3. aquitard vertical hydraulic conductivity, in gpd/sq ft
4. artesian aquifer storativity, dimensionless
5. water table aquifer storativity, dimensionless
6. aquifer thickness, in ft
7. top of aquifer depth, in ft
8. base of aquifer depth, in ft
9. initial water level depth, in ft
10. production well effective radius, in ft
11. time after pumping started at end of test, in min

The program generates drawdowns at a series of logarithmically spaced radial distances from a production well to an outer concentric boundary for a series of logarithmically spaced time increments terminating at the end of the test. Five distance intervals and five time increments are used for each tenfold increase in radius and time. In the output sections of the program, drawdowns are displayed at the production well and at five selected logarithmically spaced observation well sites at selected logarithmically spaced times, including the end of the test. Then drawdowns are displayed at the end of the test at selected logarithmically spaced distances from the production well. The program simulates infinite aquifer system conditions by setting the distance to the outer boundary equal to 1×10^5 ft beyond the influence of pumping. Wells are assumed to fully penetrate the aquifer.

These drawdowns may be algebraically added to drawdown or buildup due to the impacts of partial penetration, interference, barrier and/or recharge boundaries, and aquifer system discontinuities, thereby simulating various combinations of aquifer system and well conditions. Partial penetration impacts may be estimated with microcomputer program PT2 (provided in Appendix A) which utilizes a well function approximation presented by Sandberg et al. (1981). Partial penetration impacts may be either positive or negative, depending on well construction features.The interactive database section of the program prompts the user to enter values of the following aquifer system characteristics and well conditions:

1. drawdown with full penetration, in ft
2. production well discharge rate, in gpm
3. aquifer storativity, dimensionless
4. time after pumping started, in min
5. aquifer horizontal hydraulic conductivity, in gpd/sq ft
6. aquifer vertical hydraulic conductivity, in gpd/sq ft
7. radial distance to observation well, in ft
8. aquifer thickness, in ft
9. distance from aquifer top to bottom of production well, in ft
10. distance from aquifer top to top of production well screen, in ft
11. distance from aquifer top to bottom of observation well, in ft
12. distance from aquifer top to top of observation well screen, in ft

The output section of the program displays calculated partial penetration impact and the total drawdown with partial penetration.

Interference and boundary impacts may be estimated with the Rathod and Rushton microcomputer program mentioned earlier. Boundary impacts involve the image-well theory described by Ferris et al. (1962, pp. 144–166). The image-well theory may be summarized as follows: The effect of a barrier boundary on the drawdown in a well, as a result of pumping from another well, is the same as though the aquifer system were infinite and a like discharging well were located across the real boundary on a perpendicular thereto and at the same distance from the boundary as the real pumping well. For a recharge boundary, the principle is the same, except that the image well is assumed to be recharging the aquifer instead of pumping from it. Boundary problems are thereby simplified to consideration of an infinite aquifer system in which real and image wells operate simultaneously. In the case of an aquifer system discontinuity, the image well simulating a boundary is placed as usual. The image well strength is calculated with the following equation (Muskat, 1937):

$$Q_i = Q_p D_t \tag{2.3}$$

where $D_t = (T_p - T_d)/(T_p + T_d)$
Q_i = image-well strength, in gpm
Q_p = production well discharge rate, in gpm
T_p = aquifer transmissivity at production well, in gpd/ft
T_d = aquifer transmissivity beyond discontinuity, in gpd/ft

Analytical Equations

Before microcomputer programs may be utilized, the pumping test production well discharge rate and diameter must be specified. The discharge rate is the specific capacity of the production well multiplied by the available drawdown defined as the distance between the initial water level and the aquifer top under artesian conditions or the top of the screen in the production well under water table conditions. Specific capacity may be estimated with the following approximate empirical equations (see Driscoll, p. 1021):

For artesian aquifer systems

$$Q/s = T/2000 \tag{2.4}$$

For water table aquifer systems

$$Q/s = T/1500 \tag{2.5}$$

where Q/s = production well specific capacity with full penetration, in gpm/ft
Q = production well discharge rate, in gpm
s = production well drawdown, in ft
T = aquifer transmissivity, in gpd/ft

These equations assume that well loss is negligible and that the production well fully penetrates the aquifer. Specific capacity with partial penetration and no well loss may be estimated with the following equation (Kozeny, 1933, pp. 88–116):

$$Q/s_p = (Q/s)(L)(1 + 7(MN)^{1/2}) \tag{2.6}$$

where $L = L_s/m$
$M = r_w/(2mL)$
$N = \cos(3.14L/2)$
Q/s_p = production well specific capacity with partial penetration, in gpm/ft
L_s = length of screen in production well, in ft
m = aquifer thickness, in ft
r_w = production well radius, in ft

Although this equation assumes that the aquifer is homogeneous and that well loss is negligible, the calculated specific capacity is accurate enough for pumping test design purposes. Specific capacity calculations may be refined if necessary based on the results of the step drawdown test.

Table 2.2. **Production Well Diameters for Discharge Rates**

Discharge Rate (gpm)	Optimum Diameter (in.)
<100	6
75 - 175	8
150 - 350	10
300 - 700	12
500 - 1000	14
800 - 1800	16
1200 - 3000	20

The production well diameter must be large enough to accommodate a pump with a capacity equal to the discharge rate indicated by specific capacity calculations. Appropriate diameters for various pumping test discharges are presented in Table 2.2 (after Driscoll, 1986, p. 415).

The following equations and Table E.5 in Appendix E, relating the ratio of drawdown in the aquifer and aquitard to the factors u and u_c, may be utilized to estimate drawdown distribution in an aquitard observation well located at the same site as an aquifer observation well (Witherspoon and Neuman, 1972, p. 267):

$$u = 2693(r^2)S/(Tt) \quad (2.7)$$

$$u_c = 2693(Z^2)S_c/(P_c m_c t) \quad (2.8)$$

where r = distance from production well, in ft
S = aquifer storativity, dimensionless
T = aquifer transmissivity, in gpd/ft
t = time after pumping started, in min
Z = distance above aquifer top to bottom of aquitard observation well, in ft
S_c = aquitard storativity, dimensionless
P_c = aquitard vertical hydraulic conductivity, in gpd/sq ft
m_c = aquitard thickness, in ft

Example Problems

A medium to fine-grained sand and gravel nonleaky artesian aquifer occurs at a depth of 70 ft below land surface in Example Problem 2.1. It has a thickness of 50 ft and a medium degree of stratification. An aquifer barrier boundary is located about 1000 ft north of a potential pumping test production well site; otherwise the aquifer is infinite in areal extent for pumping test purposes. There are no nearby production wells. The initial water level is 50 ft below the land surface. Design the pumping test production and observation wells and determine the production well constant discharge rate and test duration that will lead to the evaluation of aquifer horizontal and vertical hydraulic conductivities, aquifer storativity, and effective location of the barrier boundary.

Based on representative hydraulic characteristic tables in Appendix C and information presented in the section of this chapter entitled "Generalized Design Features," initial estimates of aquifer horizontal hydraulic conductivity, P_V/P_H ratio, and artesian storativity could be 800 gpd/sq ft, $^1/_{10}$, and 5.0×10^{-4}, respectively. Transmissivity is the product of the aquifer horizontal hydraulic conductivity and the aquifer thickness, or 40,000 gpd/ft. Aquifer specific yield could be 0.1. From Equation 2.4, the specific capacity of the production well with full penetration could be 20 gpm/ft. Partially penetrating wells are required to provide data for the evaluation of aquifer vertical hydraulic conductivity. The specific capacity with a 25-ft-long screen in a production well with a 0.417-ft radius could be 15 gpm/ft according to Equation 2.6. To avoid storativity conversion, available drawdown could be limited to the difference between the initial water level and the aquifer top, or 20 ft. The potential yield of the production well with this restriction could be 300 gpm. Considering well loss, the constant production well discharge rate for the initial pretest conceptual model could be 250 gpm.

According to Table 2.2, the pumping test production well could have a 10-in. inside diameter, or a 0.417-ft radius. To ensure adequate partial penetration impacts, the production well could be screened opposite the lower half of the aquifer. Two observation wells could be located at the same site at a distance of 50 ft (aquifer thickness) east of the production well. One of these observation wells (OBS-1) could be screened between the depths of 10 and 12 ft below the aquifer top; the other observation well (OBS-2) could be screened between the depths of 40 and 42 ft below the aquifer top. There could be two additional observation wells. One of these observation wells (OBS-3), screened between the depths of 25 and 50 ft below the aquifer top, could be located at a distance of 500 ft east of the production well. The other observation well (OBS-4), screened between the depths of 25 and 50 ft below the aquifer top, could be spaced 1000 ft west of the production well.

The duration of the pumping test should be long enough to ensure that barrier boundary impacts are clear. Based on Equation 2.2, and noting that the boundary image well is 2000 ft north of the production well, the initial estimated test duration could be 270 minutes.

The database discussed above and microcomputer programs PT1 and PT2 were utilized to estimate drawdown distribution with the initial pretest conceptual model. A careful study of computation results indicates partial penetration impacts and drawdown distribution in wells are clear. However, boundary impacts are barely visible, and the test duration could be extended to at least 8 hours, or 480 minutes. No other modification of the initial pretest conceptual model seems necessary, and the pumping test design is declared valid. Computation results with full penetration are summarized in Table 2.3.

From Equation 1.1 with $r_c = 0.2$ ft, well storage capacity impacts could be appreciable for the first 2 minutes of pumping. Drawdown components in OBS-2 at the end of the test could be as follows: drawdown with full penetration and no boundary = 5.78 ft; drawdown due to partial pene-

Table 2.3. Computation Results for Example Problem 2.1

Distances (ft)	0.4	66.1	166.0	417.0	1047.5	2631.1
			Drawdown (ft)			
Time (min)						
0.6	7.15	0.57	0.05	0.00	0.00	0.00
1.5	8.22	1.16	0.26	0.01	0.00	0.00
2.3	8.63	1.49	0.44	0.02	0.00	0.00
3.7	9.01	1.82	0.68	0.06	0.00	0.00
5.9	9.36	2.15	0.95	0.15	0.00	0.00
9.3	9.71	2.48	1.23	0.28	0.01	0.00
14.7	10.05	2.81	1.54	0.47	0.03	0.00
23.3	10.39	3.14	1.85	0.70	0.07	0.00
36.9	10.72	3.47	2.17	0.96	0.15	0.00
58.5	11.06	3.80	2.50	1.25	0.29	0.01
92.8	11.39	4.13	2.82	1.55	0.47	0.03
147.0	11.72	4.46	3.15	1.86	0.70	0.07
233.0	12.05	4.79	3.48	2.18	0.96	0.15
369.3	12.38	5.12	3.81	2.50	1.25	0.29
480.0	12.58	5.32	4.01	2.70	1.43	0.39

Time After Pumping Started (min) = 480

Distance (ft)	Drawdown (ft)
0.4	12.58
17.6	7.30
26.3	6.64
41.7	5.98
66.1	5.32
104.8	4.66
166.0	4.01
263.1	3.35
417.0	2.70
660.9	2.05
1047.5	1.43
1660.1	0.85
2631.1	0.39
4170.0	0.11
6609.0	0.01

tration = 0.66 ft; drawdown due to boundary = 0.68 ft; and total drawdown = 7.12 ft.

A medium-grained sandstone leaky artesian aquifer occurs at a depth of 300 ft below land surface in Example Problem 2.2. It has a thickness of 100 ft and there are no nearby aquifer boundaries or discontinuities. A shaley sandstone aquitard with a thickness of 200 ft overlies the aquifer. The initial water level is 80 ft below the land surface. A nearby production well is located 2000 ft south of a potential pumping test site. The pump in that well is operated 8 hours per day at a rate of 100 gpm. Design the pumping test production and observation wells and determine the production well constant discharge rate and test duration that will lead to the evaluation of the aquifer horizontal hydraulic conductivity, aquifer storativity, aquitard vertical hydraulic conductivity, and storativity.

Based on representative hydraulic characteristic tables in Appendix C and information presented in the section of this chapter entitled "Generalized Design Features," initial estimates of aquifer horizontal hydraulic conductivity, aquifer artesian storativity, and aquitard vertical hydraulic conductivity could be 10 gpd/sq ft, 1.0×10^{-4}, 1.0×10^{-3}, respectively. Transmissivity could be 1000 gpd/ft and aquifer specific yield could be 0.05. Based on Equation 2.4, the specific capacity of the fully penetrating production well could be 0.5 gpm/ft. To avoid storativity conversion, available drawdown is limited to the difference between the initial water level and the aquifer top, or 220 ft. The potential yield of the production well with this restriction could be 110 gpm.

According to Table 2.2, the production well could have an 8-in. inside diameter, or a 0.333-ft radius. One fully penetrating observation well (OBS-1) could be located 80 ft west of the production well, and another fully penetrating observation well (OBS-2) could be located 800 ft east of the production well. One fully penetrating aquifer observation well (OBS-3) and one partially penetrating aquitard observation well (OBS-4) could be located at the same site 300 ft east of the production well. The aquitard observation well could be open between the depths of 45 and 55 ft above the aquifer top.

The duration of the pumping test should be long enough to ensure that drawdown in the aquitard well will be clear. A duration of 8 hours or 480 minutes could be selected for the initial pretest conceptual model to determine the nature of aquitard drawdown.

The database discussed above and microcomputer programs PT1 and Equations 2.7 and 2.8 were utilized to estimate drawdown distribution with the initial pretest conceptual model. Computation results with $S_c = 1.0 \times 10^{-4}$ indicate that drawdown in the aquitard observation well for a pumping period of 480 minutes is negligible. Otherwise, drawdown distribution with the initial pretest conceptual model seems reasonable. The initial pretest conceptual model pumping test duration was changed to 1440 minutes (1 day) and computations were repeated with the modified pretest conceptual model. These computation results are summarized in Table 2.4. From Equation 1.1 with $r_c = 0.1$ ft, well storage impacts could be appreciable for the first 54 minutes of the test.

A careful study of computation results indicates that drawdowns at the end of the test in the aquifer observation well and aquitard observation well (2 ft) are clear. Drawdown due to the nearby production well could be appreciable, indicating that interference should be controlled or eliminated. No further modifications related to the pumping test design are necessary, and the pumping test design is declared valid.

Aquitard hydraulic characteristics may be evaluated based on drawdown data for an aquitard observation well and/or leakage impact data for an aquifer observation well. Without the aquitard observation well, the test duration should be long enough to ensure that aquitard leakage impacts on water levels in the aquifer are clear. In Example Problem 2.2, without the aquitard observation well the duration of the pumping test would have to be about 100

Table 2.4. Computation Results for Example Problem 2.2

Distance (ft)	0.3	52.8	132.6	333.0	836.5	2101.1
			Drawdown (ft)			
Time (min)						
0.5	17.76	0.11	0.00	0.00	0.00	0.00
1.2	38.98	1.02	0.02	0.00	0.00	0.00
3.0	74.80	5.05	0.41	0.00	0.00	0.00
4.7	96.33	9.17	1.23	0.01	0.00	0.00
7.5	117.27	14.79	2.98	0.07	0.00	0.00
11.9	135.26	21.47	5.95	0.29	0.00	0.00
29.9	159.82	35.51	15.12	2.24	0.03	0.00
47.3	168.14	42.17	20.60	4.54	0.12	0.00
75.0	175.29	48.53	26.30	7.83	0.41	0.00
118.9	181.85	54.69	32.07	11.93	1.14	0.01
298.6	194.18	66.68	43.69	21.72	4.96	0.13
473.2	200.15	72.57	49.50	27.07	8.23	0.44
750.1	206.06	78.43	55.30	32.59	12.26	1.20
1188.8	211.91	84.26	61.09	38.20	16.86	2.67
1440.0	214.57	86.90	63.73	40.77	19.07	3.54

Time After Pumping Started (min) = 1440

Distance (ft)	Drawdown (ft)
0.3	214.57
21.0	110.11
52.8	86.90
83.6	75.31
132.6	63.73
211.0	52.19
333.0	40.77
527.8	29.60
836.5	19.07
1325.7	9.95
2101.1	3.54
3330.0	0.67
5277.7	0.05
8364.6	0.00

days. Aquitard leakage impacts in aquifer observation wells are not clear until that time.

A coarse to medium-grained sand and gravel water table aquifer occurs at a depth of 10 ft below land surface in Example Problem 2.3. It has a thickness of 100 ft and a medium degree of stratification. An aquifer discontinuity is located about 600 ft north of a potential pumping test production well site; otherwise the aquifer is infinite in areal extent for pumping test purposes. There are no nearby production wells. The initial water level is 10 ft below the land surface. Design the pumping test production and observation wells and determine the production well constant discharge rate and test duration that will lead to the evaluation of the aquifer horizontal and vertical hydraulic conductivities, aquifer specific yield, and effective location of the discontinuity.

Based on representative hydraulic characteristic tables in Appendix C and information presented in the section of this chapter entitled "Generalized Design Features," initial estimates of aquifer horizontal hydraulic conductivity, P_V/P_H ratio, and specific yield could be 2500 gpd/sq ft, $^1/_{10}$, and 0.05, respectively. Aquifer transmissivity could decrease from 250,000 gpd/ft at the production well site to 150,000 gpd/ft at the discontinuity. Aquifer artesian storativity could be 0.001. Based on Equation 2.5, the specific capacity of a production well with full penetration could be 167 gpm/ft. Partially penetrating wells could be selected to provide for evaluation of the aquifer vertical hydraulic conductivity. The specific capacity with a 50-ft-long screen in a production well having a 0.667-ft radius could be 124 gpm/ft, according to Equation 2.6. To limit dewatering to 10% of the aquifer thickness, available drawdown could be limited to 10 ft. The potential yield of the production well and the constant discharge rate for the initial pretest conceptual model with this restriction could be 1240 gpm.

According to Table 2.2, the pumping test production well could have a 16-in. outside diameter, or a 0.667-ft radius. To ensure adequate partial penetration impacts, the production well could be screened opposite the lower half of the aquifer. Equation 1.2 indicates partial penetration impacts are appreciable to a distance of 474 ft from the production well. Observation wells 1, 2, and 3 could be located at sites 100 ft east, 500 ft east, and 1000 ft west of the production well, respectively. The observation wells could be screened opposite the lower half of the aquifer.

It is desirable to obtain data after delayed gravity yield impacts are negligible, which according to Equation 1.3 is 1080 minutes. The duration of the pumping test should be long enough to ensure that discontinuity impacts are clear. Based on Equation 2.2, and noting that the discontinuity image well is 1200 ft north of the production well, the initial estimated test duration could exceed 1 day if the image well strength were equal to the production well discharge rate. However, the discontinuity image well strength, from Equation 2.3, could be 310 gpm. Because of the low image well strength, the test duration could be 3 days.

The database discussed above and microcomputer programs PT1 and PT2 were utilized to estimate drawdown distribution with the initial pretest conceptual model. The drawdown due to partial penetration in OBS-1 at the end of the test could be 0.4 ft. Computation results are summarized in Table 2.5. From Equation 1.1, with $r_c = 0.33$ ft, well storage impacts could be negligible after 1 minute of pumping. A careful study of computation results indicates partial penetration and discontinuity impacts are clear. Drawdown distribution is adequate for analysis. No modification of the initial pretest conceptual model seems necessary, and the pumping test design is declared valid.

DATA COLLECTION AND PROCESSING

Discharge and water level data collected in the field usually are recorded on forms with headings such as: Well No., Distance from Production Well (ft), Surface Water

Table 2.5. Computation Results for Example Problem 2.3

Distance (ft)	0.67	105.7	265.5	667.0	1675.4	4208.5
	Drawdown (ft)					
Time (min)						
0.1	4.53	0.07	0.00	0.00	0.00	0.00
0.5	6.10	0.47	0.04	0.00	0.00	0.00
0.8	6.45	0.67	0.09	0.00	0.00	0.00
1.2	6.74	0.88	0.17	0.00	0.00	0.00
4.8	7.45	1.46	0.54	0.05	0.00	0.00
7.6	7.64	1.63	0.67	0.08	0.00	0.00
12.0	7.80	1.77	0.79	0.13	0.00	0.00
30.2	8.07	2.02	1.01	0.23	0.00	0.00
47.9	8.19	2.13	1.10	0.28	0.01	0.00
75.9	8.29	2.22	1.19	0.34	0.01	0.00
120.4	8.40	2.32	1.28	0.40	0.02	0.00
302.3	8.61	2.51	1.47	0.54	0.04	0.00
479.2	8.73	2.63	1.57	0.62	0.06	0.00
759.5	8.87	2.76	1.70	0.73	0.10	0.00
1203.7	9.04	2.92	1.86	0.86	0.15	0.00
1907.7	9.24	3.11	2.04	1.03	0.24	0.01
3023.5	9.47	3.32	2.24	1.21	0.36	0.02
4320.0	9.66	3.50	2.42	1.38	0.47	0.03

Distance (ft)	Drawdown (ft)
0.7	9.66
26.5	5.14
66.7	4.04
105.7	3.50
265.5	2.42
420.8	1.89
667.0	1.38
1057.1	0.90
1675.4	0.47
2655.4	0.17
4208.5	0.03
6670.0	0.00

Measuring Point No., Date, Hour, Time After Pumping Started or Stopped (min), Depth to Water Level (ft), Water Level Change (ft), Discharge (gpm), and Remarks. Depths to water in wells and at surface water measurement points may be measured with a steel tape coated at the end with blue carpenter's chalk, electric tapes, float-actuated mechanical recorders, electric pressure transducers, or air lines with a pressure gage. The measuring point at each observation well should be clearly marked and its elevation determined. The circular weir is the most commonly used device to measure the rate of discharge from a turbine pump. (For other details concerning data collection see Driscoll, 1986, pp. 534-552.)

With respect to field observation, required records and the tolerance in measurement generally considered acceptable are as follows (Stallman, 1971, p. 11):

1. production well discharge (10%)
2. depth to water in wells below measuring point (0.01 ft)
3. distance from production well to each observation well (0.05%)
4. synchronous time (1% of time after pumping started or stopped)
5. descriptions of measuring points
6. elevation of measuring points (0.1 ft)
7. vertical distance between measuring point and land surface (0.1 ft)
8. total depths of wells (1%)
9. depths and lengths of screened or open intervals of wells (1%)
10. diameter, casing type, screen type, and method of well construction (nominal)
11. location of wells in plan view, relative to land survey net or by latitude and longitude (accuracy varies)
12. barometric pressure

Barometric scales may be graduated in inches (in.), millimeters (mm), or millibars (mb). Conversion from one scale to another is facilitated by using the following equation: 1 in. = 25.4 mm = 33.86 mb.

Drawdown and Recovery

To determine drawdown or recovery, the water level trend before pumping started or after pumping stopped (antecedent trend) should be extrapolated through the pumping or recovery period, and differences between extrapolated and observed water levels are calculated. The accuracy of drawdown or recovery calculations is equally dependent upon extrapolated trends and measured water levels.

Recovery is a mirror image of drawdown, provided aquifer system hydraulic characteristics and conditions have not changed and extrapolations are correct. The analyses of drawdown and recovery data should yield the same results. Drawdown or recovery is the difference between where water levels are and where water levels would be if the pumping test were not conducted. Extrapolation is often facilitated by obtaining water level data in an observation well immediately outside the area of influence of the pumping test. Erroneous pumping test results will be obtained if antecedent trends are ignored.

Water Level Adjustments

Before water level data are analyzed, they must be adjusted for any pumping rate changes in the test production well or in nearby production wells, and/or barometric pressure changes. Time rate of drawdown changes due to interference and/or barometric pressure trend changes may be erroneously interpreted as boundary impacts.

Under artesian conditions, barometric pressure change adjustments are made by selecting a time interval during which water levels are not affected by pumping rate changes. Barometric readings are inverted and plotted on plain coordinate paper, together with water level data for the production and observation wells. Prominent barometric changes expressed in feet of water (1 in. of mercury = 1.13 ft of water) are compared to corresponding water level changes. The amount of rise in water level as a result of a decrease in barometric pressure and the amount of decline in water level as a result of an increase in barometric pressure are calculated. The barometric efficiency is then calculated with the following equation (Ferris et al., 1962, p. 85):

$$BE = (W/B)100 \tag{2.9}$$

Barometric changes may exceed 1 in. of mercury (1.13 ft of water) and barometric efficiency under artesian conditions commonly exceeds 50%. Barometric efficiency commonly is negligible under water table conditions.

Drawdown data are adjusted for barometric pressure changes occurring during a pumping test using records of barometric pressure changes and the following equation:

$$W = (BE \times B)/100 \tag{2.10}$$

where BE = barometric efficiency (%)
W = change in water level (ft)
B = change in barometric pressure (ft of water)

Antecedent barometric trends should be considered in estimating changes in barometric pressure.

The tidal or river efficiency may be calculated and drawdown data may be adjusted for any surface water stage changes occurring during a pumping test by obtaining a record of surface water stage fluctuations prior to and during the test and using the following equations (Ferris et al., 1962, p. 85):

$$TE = (WS/H)100 \tag{2.11}$$

$$RE = (WS/H)100 \tag{2.12}$$

$$WS = (TE \times H)/100 \tag{2.13}$$

$$WS = (RE \times H)/100 \tag{2.14}$$

where TE = tidal efficiency (%)
RE = river efficiency (%)
WS = change in water level (ft)
H = change in surface stage (ft)

The application of heavy loads in the vicinity of some artesian wells causes changes in water levels. Fluctuations in water levels sometimes occur when railroad trains or trucks pass pumping test sites. Drawdown data must be adjusted for these changes in aquifer loading before they are used to determine aquifer system hydraulic characteristic values. Earth tides and earthquakes affect water levels and must be given due attention (see Ferris et al., 1962, pp. 86–87).

Under water table conditions, gravity drainage of interstices due to pumping may appreciably decrease the saturated aquifer thickness and, therefore, transmissivity. Analytical models for analyzing pumping test data are based on the assumption that drawdown is negligible in comparison to the initial saturated thickness of the aquifer. Thus, drawdown data must be adjusted for the effects of dewatering before they are used to determine aquifer system hydraulic characteristic values with the following equation (Jacob, 1944):

$$s_a = s_o - s_o^2/(2m) \tag{2.15}$$

where s_a = adjusted drawdown (ft)
s_o = observed drawdown (ft)
m = initial saturated aquifer thickness (ft)

Equation 2.15 is strictly applicable to late drawdown data and not to early and intermediate data (Neuman, 1975, pp. 334–335).

Data manipulation, including drawdown adjustment, may be facilitated with a microcomputer general purpose database management system such as Reflex The Analyst (trademark of Borland International,Inc.). This software allows users to view data in several ways: in single record form, in lists, in graphs, and in cross tabulations. Records may be statistically analyzed, sorted, and filtered. For example, fields (water level data and adjustments) may be added or subtracted to determine record (time) drawdown totals (see Ericson and Moskol, 1986).

3

Constant Discharge Test Analysis

Constant discharge test analysis utilizes analytical models and equations. Selection of the appropriate model is guided by the results of the pretest conceptual modeling effort. Although there are numerous available models (see Walton, 1985, pp. 141-247), six are frequently applied to pumping test analysis. A brief description of these models follows.

ANALYTICAL MODELS AND EQUATIONS

Model 1 (Theis, 1935, pp. 519–524) assumes that flow is entirely horizontal and radial and that wells fully penetrate the aquifer. There are no vertical components of flow. No water is stored within wells; therefore, drawdown or recovery data are not affected by well storage capacity. The uniformly porous aquifer is overlain and underlain by aquicludes with negligible vertical hydraulic conductivity. The nonleaky artesian aquifer is homogeneous, isotropic, infinite in areal extent, and constant in thickness throughout. Wells have infinitesimal diameters and the discharge rate is constant. There are no boundaries or discontinuities. Model 1 simulates water table conditions when delayed gravity yield impacts are negligible. The model 1 equation is:

$$s = 114.6QW(u)/T \tag{3.1}$$

where $$u = 2693r^2S/(Tt) \tag{3.2}$$

s = drawdown, in ft
Q = discharge rate, in gpm
$W(u)$ = well function, dimensionless
T = aquifer transmissivity, in gpd/ft
r = distance from production well, in ft
S = aquifer storativity, dimensionless
t = time after pumping started, in min

Model 2 (Papadopulos, 1967, pp. 241–244) is the same as model 1 except the production well has a finite diameter and storage capacity. In model 1, flow into the production well from the aquifer equals the constant flow that is pumped from the well. In model 2, flow into the production well from the aquifer equals the constant flow that is pumped from the well minus the flow from the change in storage of water in the well as the water level declines. Thus, flow from the aquifer into the production well is variable during early pumping periods (see Hunt, 1983, pp. 221–224). The model 2 equation is:

$$s = 114.6QW(u,S,Rho)/T \tag{3.3}$$

where $$Rho = r/rw \tag{3.4}$$

$W(u,S,Rho)$ = well function, dimensionless
r_w = production well effective radius, in ft

Model 3 (Hantush and Jacob, 1955, pp. 95–100) assumes wells fully penetrate a leaky artesian aquifer overlain by an aquitard and underlain by an aquiclude. Overlying the aquitard are deposits (source bed) in which there is a water table. The aquifer is homogeneous, isotropic, infinite in areal extent, and constant in thickness throughout. Flow lines are assumed to be refracted a full right angle as they cross the aquitard-aquifer interface. The aquitard is assumed to be more or less incompressible so that water released from storage therein is negligible. Drawdown in the source bed is negligible, production and observation wells have infinitesimal diameters and no storage capacity, and the discharge rate is constant. There are no boundaries or discontinuities. Neuman and Witherspoon (1969, pp. 810 and 821) indicate that the use of model 3 is justified when $r/(4m)(P_cS_cm/(P_HSm_c))^{1/2} < 0.1$ and $t < 1.08 \times 10^3\ m_cS_c/P_c$. Under these conditions, change in storage of water in the aquitard and drawdown in the source bed are negligible. The model 3 equation is:

$$s = 114.6QW(u,r/B)/T \qquad (3.5)$$

where

$$B = (Tm_c/P_c)^{1/2} \qquad (3.6)$$

$W(u,r/B)$ = well function, dimensionless
m_c = aquitard thickness, in ft
P_c = aquitard vertical hydraulic conductivity, in gpd/sq ft

Model 4 (Hantush, 1964, pp. 335–336) is the same as model 3, except that water is released from storage within the aquitard and it is assumed that the effects of pumping have not reached the aquitard top. Hantush (1960, p. 3716) indicates that the use of model 4 is justified when source bed drawdown is negligible, that is, $t < 1.08 \times 10^3\ m_cS_c/P_c$. Other models are available for analyzing data for larger values of time. The model 4 equation is:

$$s = 114.6QW(u,Gamma)/T \qquad (3.7)$$

where

$$Gamma = (r/4)[S_cP_c/(TSm_c)]^{1/2} \qquad (3.8)$$

$W(u,Gamma)$ = well function, dimensionless
S_c = aquitard storativity, dimensionless

Model 5 (Witherspoon and Neuman, 1972; p. 267) is the same as model 4 except there is a partially penetrating aquitard observation well. Model Equations 2.7 and 2.8

and Table E.5 in Appendix E, relating the ratio of drawdowns in the aquifer and aquitard at the same site to u and u_c, were discussed previously.

Model 6 (Neuman, 1975, pp. 329–342) assumes wells fully penetrate a uniformly porous water table aquifer underlain by an aquiclude. The aquifer is homogeneous, anisotropic (stratified), infinite in areal extent, and constant in thickness throughout. Drawdown is negligible in comparison to the initial saturated aquifer thickness. Principal aquifer hydraulic conductivities are oriented parallel to the coordinate axes. Wells have infinitesimal diameters and no storage capacity, and the discharge rate is constant. There are no boundaries or discontinuities. The model 6 equation is:

$$s = 114.6QW(u_A,u_B,Beta)/T \qquad (3.9)$$

where

$$u_A = 2693r^2S/(Tt) \qquad (3.10)$$

$$u_B = 2693r^2S_y/(Tt) \qquad (3.11)$$

$$Beta = r^2P_V/(m^2P_H) \qquad (3.12)$$

$W(u_A,u_B,Beta)$ = well function, dimensionless
S_y = aquifer specific yield, dimensionless
P_V = aquifer vertical hydraulic conductivity, in gpd/sq ft
P_H = aquifer horizontal hydraulic conductivity, in gpd/sq ft

Model equations may be modified with the image well theory to simulate finite (bounded) aquifer systems. The well function for an observation well in an aquifer having a single boundary is the algebraic summation of the production well function and the image well function (see Stallman, 1963, p. 46). For example, model 1 with a single boundary will have a total well function and drawdown as follows, depending upon whether a barrier or recharge boundary is involved:

$$W_t = W(u) + W(u_i) \text{ or } W(u) - W(u_i) \tag{3.13}$$

$$s_t = s + s_i \text{ or } s - s_i \tag{3.14}$$

where

$$s_i = 114.6QW(u_i)/T \tag{3.15}$$

$$u_i = 2693r_i^2S/(Tt) \tag{3.16}$$

W_t = total well function, dimensionless
$W(u_i)$ = image well function, dimensionless
$W(u)$ = production well function, dimensionless
r_i = distance between observation and image wells, in ft
s_t = total drawdown, in ft
s = drawdown due to production well, in ft
s_i = drawdown or buildup due to image well, in ft

Model equations also may be modified to simulate well partial penetration conditions. The well function for a partially penetrating well is the algebraic summation of the well function with fully penetrating conditions and a partial penetration impact well function (see Hantush, 1964; pp. 347-358). For example, the total well function and drawdown equations for model 1 with partially penetrating wells are as follows:

$$W_t = W(u) + W(u,r(P_V/P_H)\hat{}0.5/m,l/m,d/m,l_o/m,d_o/m) \tag{3.17}$$

$$s_t = s + s_p \tag{3.18}$$

where

$$s_p = 114.6QW[u,r(P_V/P_H)\hat{}0.5/m,l/m,d/m,l_o/m,d_o/m]/T \tag{3.19}$$

$W(u,r(P_V/P_H)\hat{}0.5/m,l/m,d/m,l_o/m,d_o/m)$
= partial penetration impact well function, dimensionless
l = distance from aquifer top to production well base, in ft
m = aquifer thickness, in ft
d = distance from aquifer top to production well screen top, in ft

l_o = distance from aquifer top to observation well base, in ft
d_o = distance from aquifer top to observation well screen top, in ft
s_p = partial penetration impact, in ft

Under finite aquifer and partial penetration conditions, the total well function is the algebraic summation of the production well function with infinite aquifer and fully penetrating well conditions, the image well function, and the partial penetration impact well function.

Well functions for models 1, 3, 4, 5, and 6 may be adjusted to include well storage capacity impacts with model 2 type curve patches for small values of 1/u consistent with Equation 1.1.

Hantush (1966, pp. 421–426) provides equations for an anisotropic model aquifer system in which the hydraulic conductivity varies in different directions. The x-direction is taken to be parallel to the major axis of anisotropy and the y-direction lies along the minor axis.

Well Function Microcomputer Programs

Model well function (for example, W[u]) values, over the practical range of the dimensionless variable (for example, u) values, are listed in Appendix E. Interpolation between variable values listed may be rapidly accomplished with microcomputer program PT3 provided in Appendix A. The interactive database input section of the program prompts the user to enter variable and well function values for six points surrounding the interpolation point and the variable value for the interpolation point. The output section of the program displays the estimated well function value for the interpolation point.

Microcomputer programs PT2, PT4, and PT5, listed in Appendix A, generate values of $W(u,r(P_V/P_H)$ ^0.5/m, l/m, d/m, l_o/m, d_o/m), W(u), and W(u,r/B), respectively. These programs are modifications of programs presented by Walton (1985, pp. 341–342) and are based on polynomial approximations of well function equations (see Walton, 1985, pp. 337–340). Listings of FORTRAN programs for calculating values of W(u), W(u,r/B), W(u,S,Rho), and W(u,Gamma) are provided by Reed (1980, pp. 57–106). Listing of a FORTRAN program for calculating values of W(u_A,u_B,Beta) is provided by Neuman (1975).

Example Problem

The P_V/P_H ratio of a stratified nonleaky artesian aquifer is 50/500 in Example Problem 3.1. An observation well is located 35 ft from a production well. The aquifer has a thickness of 50 ft. Wells partially penetrate the aquifer and have the following construction features: depth below aquifer top to production well base = 50 ft, depth below aquifer top to production well screen top = 20 ft, depth below aquifer top to observation well base = 23 ft, and depth below aquifer top to observation well screen top = 20 ft. Prepare a table of values of W_t(u) versus u for the specified partial penetration conditions.

Values of W(u) were added algebraically to values of W(u,r(P_V/P_H)^0.5/m, l/m, d/m, l_o/m, d_o/m) with microcomputer programs PT2 and PT4, to obtain the values of W_t(u) listed in Table 3.1.

TYPE CURVE MATCHING TECHNIQUE

Aquifer system hydraulic characteristics cannot be directly determined with model equations and pumping test data because aquifer transmissivity occurs in the well function argument and again as a divisor of the well func-

Table 3.1. Computation Results for Example Problem 3.1

u	W(u)	WUPAR[a]	$W_t(u)$
0.0001	8.6332	-0.1650	8.4682
0.001	6.3315	-0.1650	6.1665
0.01	4.0379	-0.1650	3.8729
0.1	1.8229	-0.1409	1.6820
1	0.2194	-0.0012	0.2182
5	0.0011	-0.0001	0.0010

[a] WUPAR = $W(u,r(P_V/P_H)$^0.5/m,l/m,d/m,l_o/m,d_o/m)

tion. However, Theis (Wenzel, 1942, p. 88) devised a convenient type curve graphical method of superposition to analyze pumping test data. Briefly, that method involves matching logarithmic time-drawdown or distance-drawdown graphs to theoretical logarithmic well function curves (type curves) for appropriate models. Thus, type curves such as W(u) versus 1/u or W(u) versus u are matched to pumping test data plots of drawdown (s) versus time (t), s versus distance squared (r^2), or s versus r^2/t.

Type curves for models 1 through 4 and 6, plotted with logarithmic scales of 0.75 in. per log-cycle, are presented in Appendix F. Commonly, type curves are plotted with logarithmic scales of 1.85 in. per log-cycle. With ordinary graph paper, a light table is generally required for matching type curves and pumping test data plots. This inconvenience may be eliminated by printing type curves on transparent plastic, such as 0.020-in. clear acetate (see Walton, 1963, pp. 1–3).

The pretest conceptual model guides the selection of the appropriate type curve for pumping test analysis. With fully penetrating wells, one set of type curves may be applied to all observation wells. With partially penetrating wells, a special set of type curves must be generated for each observation well. Match point values are substituted into model equations for calculations of aquifer system hydraulic characteristics (see Ferris et al., 1962, p. 94).

To illustrate the type curve matching technique, consider

the single type curve associated with model 1. Values of W(u) are plotted against values of 1/u on logarithmic paper to describe a W(u) function type curve. Values of adjusted drawdown (s) are plotted on logarithmic paper of the same scale used in preparing the type curve against values of time after pumping started (t) to describe a time-drawdown curve. W(u) is related to 1/u in the same manner that s is related to t; thus, the time-drawdown curve is analogous to the type curve.

The type curve is superposed over the time-drawdown curve, keeping the W(u) axis parallel to the s axis and the 1/u axis parallel to the t axis. In the matched position, a common match point for the two curves is chosen, and the four coordinates W(u), 1/u, s, and t are recorded. For convenience, the match point may be chosen at the intersection of the major axes of the type curve. Match point coordinates are substituted into model 1 equations to calculate hydraulic characteristics. Transmissivity is calculated first and then storativity.

Interpretations of pumping test data based solely on time-drawdown data are weak. Distance-drawdown data complement time-drawdown data and should be analyzed whenever possible to strengthen interpretations. W(u) is related to u in the same manner that s is related to the square of the distances from the production well to the observation wells (r^2) for a selected time. Thus, the s versus r^2 distance-drawdown curve is analogous to the W(u) versus u type curve. The distance-drawdown curve is matched to the type curve, and hydraulic characteristics are calculated with match point coordinates. Plots of s versus t/r^2 may be matched to the W(u) versus 1/u type curve, or plots of s versus r^2/t may be matched to the W(u) versus u type curve when several observation wells are spaced at unequal distances from the production well (see Ferris et al., 1962, p. 94).

Consider model 1 with two barrier boundaries simulated with two discharging image wells. The production and image wells operate simultaneously and at the same discharge rate. The W(u) versus 1/u type curve is matched to the early portion of the time-drawdown curve for an observation well unaffected by the image well, and hydraulic characteristics are calculated as described earlier. The type curve is moved up and to the right and matched to later data affected by the first image well. The correctness of the match position is judged by noting that the s value underlying a selected W(u) value in the early data unaffected by the image well is ½ the s value underlying that same selected W(u) value in later data affected by the image well. The difference between the first type curve extrapolated trace and the second type curve trace (s_{i1}) is determined for a selected time (t_i). The hydraulic characteristics calculated with early data, the production well discharge rate, and values of s_{i1} and t_i are substituted into model 1 equations to determine the distance from the observation well to the image well (r_{i1}). The type curve is moved further up and to the right and matched to late data affected by both image wells. The correctness of the match position is judged by noting that the s value underlying a selected W(u) value in the early data unaffected by both image wells is one-third the s value underlying that same selected W(u) value in late data affected by both image wells. The difference between the second type curve extrapolated trace and the third type curve trace (s_{i2}) is determined for a selected time (t_i). The hydraulic characteristics calculated with early data, the production well discharge, and values of s_{i2} and t_i are substituted into model 1 equations to determine the distance from the observation well to the second image well (r_{i2}). In the case of partially penetrating wells, observed drawdowns must be adjusted for partial penetration impacts to obtain drawdowns that would occur under fully penetrating conditions before multiple image well analysis is performed.

Consider the family of type curves for model 3. Values of W(u,r/B) are plotted against values of 1/u on logarithmic paper for various values of r/B. Values of s plotted on logarithmic paper of the same scale as the type curve scale against values of t describe a time-drawdown curve that is analogous to one of the family of type curves.

The family of type curves is superposed on the time-drawdown curve, keeping the W(u,r/B) axis parallel with the s axis and the 1/u axis parallel to the t axis. A particular type curve is selected as being analogous to the time-drawdown curve. In the matched position a common match point is selected and coordinates W(u,r/B), 1/u, s, and t are recorded. Coordinates are used to determine the aquifer transmissivity and storativity. The value of r/B used to prepare the particular type curve found to be analogous to the time-drawdown curve is substituted into model 3 equations to determine the aquitard vertical hydraulic conductivity.

Model 6 equations describe two asymptotic families of type curves labeled $W(u_A,Beta)$ versus $1/u_A$ (type A) and $W(u_B,Beta)$ versus $1/u_B$ (type B) (Neuman, 1975, pp. 330–331) which are presented in Figures F.6 and F.7, respectively, of Appendix F. Both families of type curves approach a set of horizontal asymptotes the lengths of which depend upon the ratio sigma = S/S_y (Neuman, 1975, pp. 330–331).

The type A family of curves is superposed on early time-drawdown data, keeping the $W(u_A,Beta)$ axis parallel to the s axis and the $1/u_A$ axis parallel to the t axis. A particular type curve with a Beta value is selected as analogous to early time-drawdown data. A common match point is selected and coordinates are recorded. Coordinates are substituted into model 6 equations and the aquifer transmissivity and artesian storativity are calculated. The value of Beta is substituted into Equation 3.12 to determine the P_V/P_H ratio. The type B curve with a value of Beta previously selected is matched to intermediate and late time-drawdown data. A common match point is selected and coordinates are recorded. Coordinates are substituted into model 6 equations to calculate aquifer transmissivity and water table storativity (specific yield). Since the value of Beta used in matching both type A and type B curves is the same, calculation of the ratio P_V/P_H need not be repeated. The calculated values of aquifer transmissivities for early and late time-drawdown data should be the same. Another way to analyze time-drawdown data is to superpose both type A and type B curves on the time-drawdown curve at the same time (Neuman, 1975, pp. 330–331).

With limited and imperfect field data, the point of beginning in type curve matching is to assume that the earliest few time-drawdown values are not affected by aquitard leakage or delayed gravity yield. Analysis of intermediate and late time-drawdown data may dictate that this initial assumption be revised.

Type curve, time-drawdown, and distance-drawdown graphs may be rapidly displayed with dot-matrix printers or plotters utilizing general purpose chart or graphic microcomputer programs such as Grapher (trademark of Golden Software, Inc.). That program supports linear-linear, log-linear, linear-log, and log-log graphs; superimposition of graphs; and linear, logarithmic, exponential, power, and cubic spline curve fitting.

Pumping test data may be analyzed with the following method of successive approximations in cases where the complexities of aquifer system conditions greatly exceed those assumed in type curve models. A subjective set of hydraulic characteristics and boundary conditions is used as the database for a numerical model microcomputer program simulating well storage capacity, aquitard storativity, and delayed gravity yield. Usually, the database reflects available geological information and past experience with aquifer system response to pumping. Drawdowns calculated with the numerical model are compared with measured values of drawdown. The procedure is repeated for selected databases until calculated and observed values of drawdown match and the database is declared valid. The uniqueness of the solution is defended based on the reasonableness of the database in light of hydrogeologic information and sensitivity analyses. Time-drawdown curves generated with numerical models are analogous to complex well function type curves.

Match Point Equations

Model equations may be rearranged and expressed in terms of match point coordinates for use with type curves.

Model 1 match point equations are:

for time-drawdown and distance-drawdown curves

$$T = 114.6QW(u)_{mp}/s_{mp} \quad (3.20)$$

for time-drawdown curve

$$S = Tt_{mp}/[2693r^2(1/u)_{mp}] \quad (3.21)$$

for distance-drawdown curve

$$S = T(u)_{mp}t/[2693(r^2)_{mp}] \quad (3.22)$$

where T = aquifer transmissivity, in gpd/ft
Q = production well discharge rate, in gpm
$W(u)_{mp}$ = W(u) match point coordinate, dimensionless
s_{mp} = drawdown match point coordinate, in ft
S = aquifer storativity, dimensionless
t_{mp} = time match point coordinate, in min
r = distance from production well, in ft
$(1/u)_{mp}$ = 1/u match point coordinate, dimensionless
$(u)_{mp}$ = u match point coordinate, dimensionless
t = time after pumping started, in min
$(r^2)_{mp}$ = r^2 match point coordinate, in sq ft

Model 2 match point equations are:

$$T = 114.6QW(u,S,Rho)_{mp}/s_{mp} \quad (3.23)$$

$$S = Tt_{mp}/[2693r^2(1/u)_{mp}] \quad (3.24)$$

$$Rho = r/r_w$$

where T = aquifer transmissivity, in gpd/ft
Q = production well discharge rate, in gpm
$W(u,S,Rho)_{mp}$ = W(u,S,Rho) match point coordinate, dimensionless
s_{mp} = drawdown match point coordinate, in ft
t_{mp} = time match point coordinate, in min
r = distance from production well, in ft
$(1/u)_{mp}$ = 1/u match point coordinate, dimensionless
r_w = production well effective radius, in ft
S = aquifer storativity, dimensionless

Model 3 match point equations are:

$$T = 114.6QW(u,r/B)_{mp}/s_{mp} \tag{3.25}$$

$$S = Tt_{mp}/[2693r^2(1/u)_{mp}] \tag{3.26}$$

$$P_c = Tm_c(r/B)^2{}_{mp}/r^2 \tag{3.27}$$

where T = aquifer transmissivity, in gpd/ft
Q = production well discharge rate, in gpm
$W(u,r/B)_{mp}$ = W(u,r/B) match point coordinate, dimensionless
s_{mp} = drawdown match point coordinate, in ft
S = aquifer storativity, dimensionless
t_{mp} = time match point coordinate, in min
r = distance from production well, in ft
$(1/u)_{mp}$ = 1/u match point coordinate, dimensionless
P_c = aquitard vertical hydraulic conductivity, in gpd/ sq ft
m_c = aquitard thickness, in ft
$(r/B)_{mp}$ = r/B value of type curve found to be analogous to time-drawdown curve, dimensionless

Model 4 match point equations are:

$$T = 114.6QW(u,Gamma)_{mp}/s_{mp} \quad (3.28)$$

$$S = Tt_{mp}/[2693r^2(1/u)_{mp}] \quad (3.29)$$

$$P_c = Gamma^2{}_{mp}16TSm_c/(r^2S_c) \quad (3.30)$$

where
T = aquifer transmissivity, in gpd/ft
Q = production well discharge rate, in gpm
$W(u,Gamma)_{mp}$ = W(u,Gamma) match point coordinate, dimensionless
s_{mp} = drawdown match point coordinate, in ft
S = aquifer storativity, dimensionless
t_{mp} = time match point coordinate, dimensionless
r = distance from production well, in ft
$(1/u)_{mp}$ = 1/u match point coordinate, dimensionless
P_c = aquitard vertical hydraulic conductivity, in gpd/sq ft
$Gamma_{mp}$ = Gamma value of type curve found to be analogous to time-drawdown curve, dimensionless
m_c = aquitard thickness, in ft
S_c = aquitard storativity, dimensionless

Model 5 equations are 2.7 and 2.8 and the following (Witherspoon and Neuman, 1972, p. 267):

$$P_c = 2693Z^2(1/u_c)_{ta}S_c/(tm_c) \quad (3.31)$$

where
P_c = aquitard vertical hydraulic conductivity, in gpd/sq ft
Z = vertical distance from aquifer top to base of aquitard observation well, in ft
$(1/u_c)_{ta}$ = $1/u_c$ value corresponding to observed value

of s_c/s and calculated value of u, dimensionless

S_c = aquitard storativity, dimensionless
t = time after pumping started, in min
m_c = aquitard thickness, in ft

Model 6 match point coordinate equations are:

$$T = 114.6QW(u_A,u_B,Beta)_{mp}/s_{mp} \quad (3.32)$$

for small values of time

$$S = Tt_{mp}/[2693r^2(1/u_A)_{mp}] \quad (3.33)$$

for large values of time

$$S_y = Tt_{mp}/[2693r^2(1/u_B)_{mp}] \quad (3.34)$$

for small and large values of time

$$P_V/P_H = m^2Beta_{mp}/r^2 \quad (3.35)$$

where
T = aquifer transmissivity, in gpd/ft
Q = production well discharge rate, in gpm
$W(u_A,u_B,Beta)_{mp}$ = $W(u_A,u_B,Beta)$ match point coordinate, dimensionless
s_{mp} = drawdown match point coordinate, dimensionless
S = aquifer storativity, dimensionless
t_{mp} = time match point coordinate, in min
r = distance from production well, in ft
$(1/u_A)_{mp}$ = $1/u_A$ match point coordinate, dimensionless
$(1/u_B)_{mp}$ = $1/u_B$ match point coordinate, dimensionless
P_V = aquifer vertical hydraulic conductivity, in gpd/sq ft

P_H = aquifer horizontal hydraulic conductivity, in gpd/sq ft
m = aquifer thickness, in ft
$Beta_{mp}$ = Beta value of type curve found to be analogous to both early and late time-drawdown curves, dimensionless

The match point equations given above assume fully penetrating wells and an infinite aquifer system. If there are partial penetrating wells, appropriate partial penetration well functions must be added algebraically to full penetration well functions and total well functions should be used in match point equations.

With a single boundary aquifer system, model 1 equations to determine the distance from an observation well to an image well are:

$$W(u_i) = Ts_{de}/(114.6Q) \quad (3.36)$$

$$r_i = [Ttu_i/(2693S)]^{1/2} \quad (3.37)$$

where $W(u_i)$ = image well well function, dimensionless
T = aquifer transmissivity, in gpd/ft
s_{de} = deviation between type curve traces with and without boundary image well impacts, in ft
Q = production well discharge rate, in gpm
r_i = distance from observation well to image well, in ft
u_i = value of u_i corresponding to the calculated value of $W(u_i)$, dimensionless
S = aquifer storativity, dimensionless

When $W(u_i)$ is known, u_i is obtained from Table E.1 in Appendix E and r_i is calculated.

Example Problems

A 24-hour pumping test was conducted with wells penetrating a nonleaky artesian aquifer system in Example Problem 3.2. Observation well OBS-1 was located 20 ft west of a production well having a radius of 0.33 ft, observation well OBS-2 was located 200 ft west of the production well, and observation well OBS-3 was located 1000 ft east of the production well. A barrier boundary occurs between 700 and 1000 ft north of the production well. The aquifer thickness is 35 ft, aquifer hydraulic conductivity is moderately high, and aquifer degree of stratification is medium. The production well constant discharge rate is 150 gpm.

The production well is open between aquifer depths of 20 and 35 ft; observation wells are open between aquifer depths of 33 and 35 ft. Time-drawdown data for the observation wells, adjusted for barometric pressure changes, are presented in Table 3.2. Calculate aquifer transmissivity and storativity, distances from observation wells to the barrier boundary image well, and the degree of aquifer stratification (P_V/P_H).

The pretest conceptual modeling effort indicates partial penetration impacts are negligible in OBS-2 and OBS-3; partial penetration impacts are appreciable in OBS-1. The W(u) versus 1/u type curve in Figure F.1 of Appendix F was matched to early portions of the time-drawdown curves for OBS-2 and OBS-3 unlikely to be affected by the barrier boundary. Match point coordinates for OBS-2 are: W(u) = 1.0, 1/u = 10, s = 0.49 ft, and t = 12 min. Match point coordinates for OBS-3 are: W(u) = 1.0, 1/u = 10, s = 0.49 ft, and t = 305 min. Values of transmissivity and storativity calculated with match point coordinates and Equations 3.20 and 3.21 are 35,082 gpd/ft and 0.00039, respectively. These values and Equation 1.1 with r_c = 0.1 ft indicate drawdown for time periods of 1 and 2 minutes fall below the type curves traces because of well storage capacity

Table 3.2. Database for Example Problem 3.2

Time after Pumping Started (min)	OBS-1 Adjusted Drawdown (ft)	OBS-2 Adjusted Drawdown (ft)	OBS-3 Adjusted Drawdown (ft)
1	2.60	.07	
2	3.15	.21	
3	3.50	.33	
4	3.63	.43	
5	3.75	.51	
6	3.88	.58	
7	3.97	.65	
8	4.02	.70	
9	4.12	.74	
10	4.21	.80	.01
20	4.52	1.08	.05
30	4.61	1.26	.11
40	4.83	1.40	.16
50	5.00	1.51	.22
60	5.17	1.62	.28
70	5.36	1.73	.33
80	5.42	1.81	.37
90	5.53	1.90	.43
100	5.64	1.97	.48
200	6.21	2.45	.91
300	6.50	2.77	1.22
400	6.82	3.06	1.43
500	6.97	3.24	1.65
600	7.02	3.50	1.72
700	7.05	3.61	1.95
800	7.18	3.82	2.06
900	7.23	3.96	2.17
1000	7.31	4.02	2.22
1440	7.72	4.45	2.58

impacts. According to Equation 1.2, partial penetration impacts are negligible beyond a distance of 166 ft and OBS-1.

A family of $W_t(u)$ versus 1/u type curves for OBS-1 were prepared with microcomputer program PT2 for three selected values of P_V/P_H. The type curve with a P_V/P_H value of $^1/_{10}$ was found to be analogous to the time-drawdown curve. Curve matching was guided by previously calculated values of transmissivity and storativity. Match point coordinates are: $W_t(u) = 1.0$, $1/u = 10$, $s = 0.49$ ft, and $t = 0.12$ min.

After about 70 minutes of pumping, observation well drawdowns deviate from type curve traces due to barrier boundary impacts. Deviations for a time period of 1000 minutes near the end of the test and Equations 3.36 and 3.37 were utilized to determine the distances from observation wells to the boundary image well. $s_{i1} = 1.05$ ft, $s_{i2} = 1.03$ ft, and $s_{i3} = 0.90$ ft. $W(u_{i1}) = 2.1429$, $W(u_{i2}) = 2.1021$, and $W(u_{i3}) = 1.8368$, $u_{i1} = 7.05 \times 10^{-2}$, $u_{i2} = 7.35 \times 10^{-2}$, and $u_{i3} = 9.8 \times 10^{-2}$. Distances from the image well to OBS-1, OBS-2, and OBS-3 are 1842, 1880, and 2171 ft, respectively. Thus, the effective barrier boundary is located about 900 ft north of the production well.

The partial penetration impact (s_p) in OBS-1 for a time period of 50 minutes before boundary impacts are appreciable was estimated with the selected P_V/P_H ratio, calculated transmissivity and storativity, and Equation 3.19. This impact (1.20 ft) was subtracted from the adjusted drawdown (s_t) for the same time period to determine the drawdown (3.80 ft) that would be observed under fully penetrating conditions (s). This value of s and drawdowns for a time period of 50 minutes in OBS-2 and OBS-3 were plotted against corresponding values of r^2 to describe a distance-drawdown curve. The W(u) versus u type curve in Figure F.2 of Appendix F was matched to the distance-drawdown curve and transmissivity and storativity values were calculated with match point coordinates and Equations 3.20 and 3.22. These values are the same as those calculated with time-drawdown curves and the problem solution is declared valid. Well function, time-drawdown, and distance-drawdown graphs are presented in Figures 3.1, 3.2, and 3.3.

An 8-hour pumping test was conducted with a production well and one observation well in a nonleaky artesian

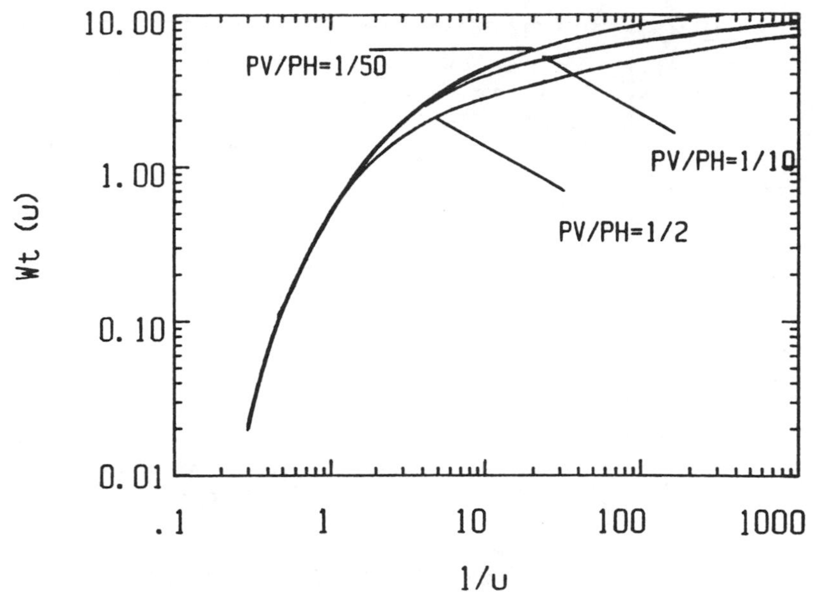

Figure 3.1. Well function graph for Example Problem 3.2.

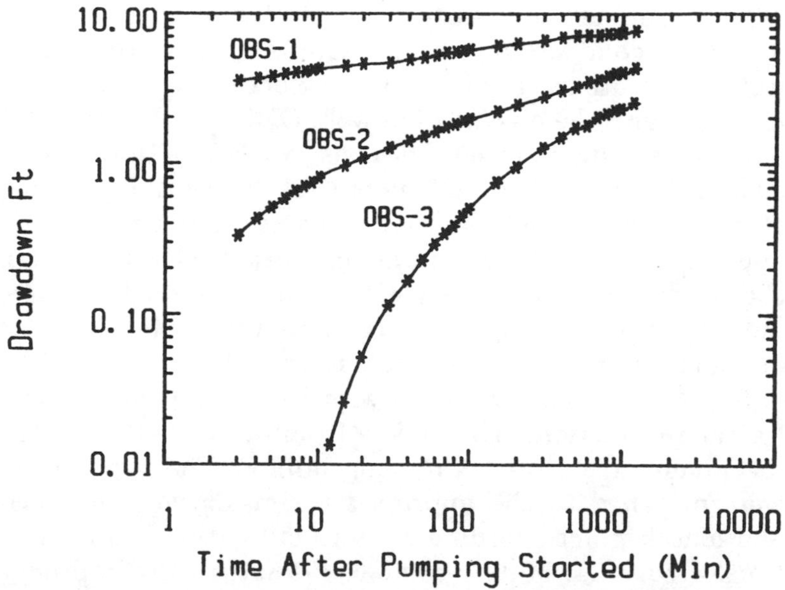

Figure 3.2. Time-drawdown graphs for Example Problem 3.2.

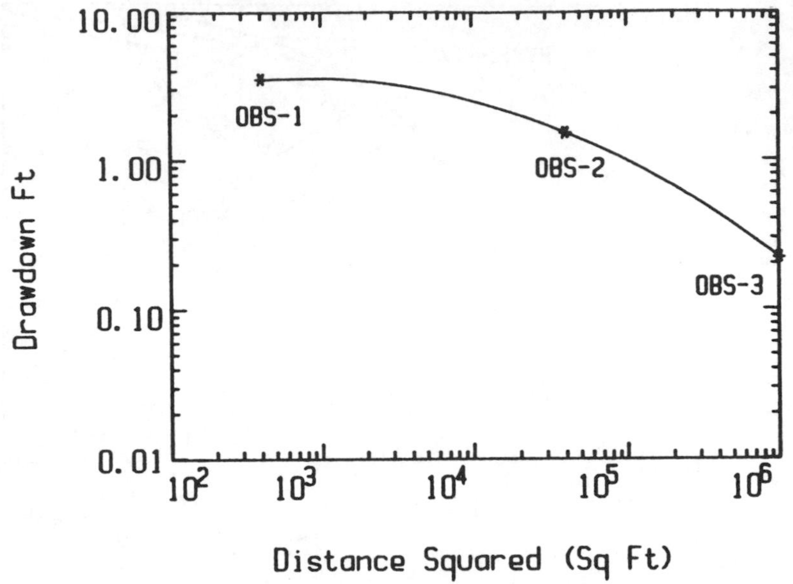

Figure 3.3. Distance-drawdown graph for Example Problem 3.2.

aquifer system which is 100 ft thick in Example Problem 3.3. There are no nearby aquifer boundaries and the hydraulic conductivity of the aquifer is low. Both wells fully penetrate the aquifer. The production well has a 0.25-ft radius and the observation well (OBS-1) is located 25 ft from the production well. The constant production well discharge rate was 10 gpm. The pretest conceptual modeling effort indicates well storage capacity impacts may be appreciable during early pumping periods. Drawdowns in the production and observation wells, adjusted for barometric pressure changes, are presented in Table 3.3. Calculate aquifer transmissivity and storativity.

The W(u,S,Rho) versus 1/u with S = 0.0001 type curves for the production well (Rho = 1) and OBS-1 (Rho = 100) are given in Figure F.3 of Appendix F. The type curves were matched to the appropriate time-drawdown curves and match point coordinates were substituted into Equations 3.23–3.24. Match point coordinates for the production well are: W(u,S,Rho) = 1.0, 1/u = 10, s = 11 ft, and t =

Table 3.3. Database for Example Problem 3.3

Time After Pumping Started (min)	Production Well Adjusted Drawdown (ft)	OBS-1 Adjusted Drawdown (ft)
4	25.23	1.01
5	30.67	1.71
6	36.12	2.36
7	41.25	3.19
8	46.14	3.92
9	51.06	4.63
10	55.11	5.39
20	82.41	12.41
30	100.02	17.79
40	113.33	22.88
50	122.46	25.67
60	129.89	28.54
70	136.22	30.82
80	139.79	33.61
90	144.17	34.97
100	146.62	36.83
200	160.12	48.11
300	169.86	53.56
480	177.73	61.02

= 1.6 min. Match point coordinates for OBS-1 are: W(u,S,Rho) = 1.0, 1/u = 10, s = 11 ft, and t = 16.5 ft. Time-drawdown graphs for the production well and observation well 1 are presented in Figure 3.4.

Calculated values of transmissivity and storativity are 104 gpd/ft and 0.0001, respectively. Equation 1.1 with the calculated value of transmissivity and r_c = 0.05 ft indicates well storage capacity impacts are appreciable until about 324 minutes after pumping started. The first few minutes of time-drawdown data describe a straight line indicating that initially most of the water discharged from the production well was taken from storage within the production well itself.

A three-day pumping test was conducted with fully penetrating wells in a leaky artesian aquifer system having no

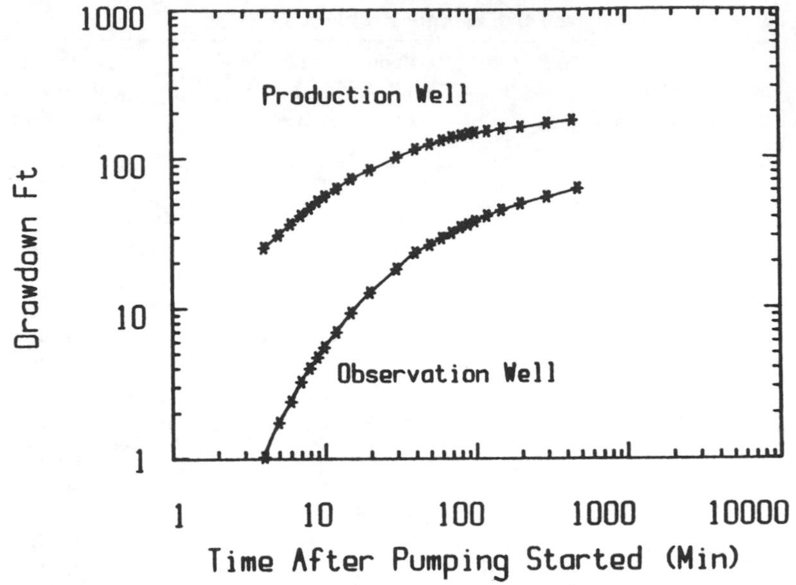

Figure 3.4. Time-drawdown graphs for Example Problem 3.3.

nearby boundaries in Example Problem 3.4. Aquifer and aquitard thicknesses are 50 and 25 ft, respectively. The aquifer hydraulic conductivity is low and the aquitard vertical hydraulic conductivity is high. The pretest conceptual modeling effort indicates aquitard storativity impacts are negligible. The production well radius and constant discharge rate are 0.33 ft and 100 gpm, respectively. Time-drawdown data, adjusted for barometric pressure changes, for an observation well (OBS-1) located 75 ft from the production well are presented in Table 3.4. Calculate aquifer transmissivity and storativity and aquitard vertical hydraulic conductivity.

A family of W(u,r/B) versus 1/u type curves for selected values of r/B are presented in Figure F.4 of Appendix F. The time-drawdown curve for OBS-1 was found to be analogous to the r/B = 0.03 type curve. Match point coordinates and the r/B value of 0.03 were substituted into Equations 3.25–3.27. Match point coordinates are: W(u,r/B) = 1.0, 1/u = 10, s = 0.46 ft, and t = 2.4 min. Calculated aquifer

Table 3.4. Database for Example Problem 3.4

Time After Pumping Started (min)	OBS-1 Adjusted Drawdown (ft)	Time After Pumping Started (min)	OBS-1 Adjusted Drawdown (ft)
1	.49	70	2.42
2	.78	80	2.46
3	.93	90	2.51
4	1.08	100	2.52
5	1.17	200	2.69
6	1.25	300	2.92
7	1.31	400	2.98
8	1.37	500	3.02
9	1.41	600	3.04
10	1.46	700	3.07
20	1.78	800	3.10
30	1.97	900	3.11
40	2.16	1000	3.12
50	2.28	2000	3.26
60	2.33	4320	3.31

transmissivity and storativity and aquitard vertical hydraulic conductivity are 24,913 gpd/ft, 0.00039, and 0.1 gpd/sq ft, respectively. Leakage impacts are negligible until about 300 minutes after pumping started and are clear at the end of the test. The time-drawdown graph for OBS-1 is presented in Figure 3.5.

A 2-day pumping test was conducted with wells penetrating a leaky artesian aquifer system in Example Problem 3.5. The production well, with a radius of 0.67 ft, was pumped at a constant rate of 1000 gpm and drawdowns were measured in an aquifer observation well (OBS-1) and in an aquitard observation well (OBS-2) located at the same site 100 ft from the production well. Aquifer wells fully penetrate the aquifer; the aquitard observation well is open in a short interval at a distance of 10 ft above the aquifer top. The aquifer and aquitard thicknesses are 80 and 32 ft, respectively. There are no nearby aquifer boundaries or interfering production wells; the aquifer hydraulic conduc-

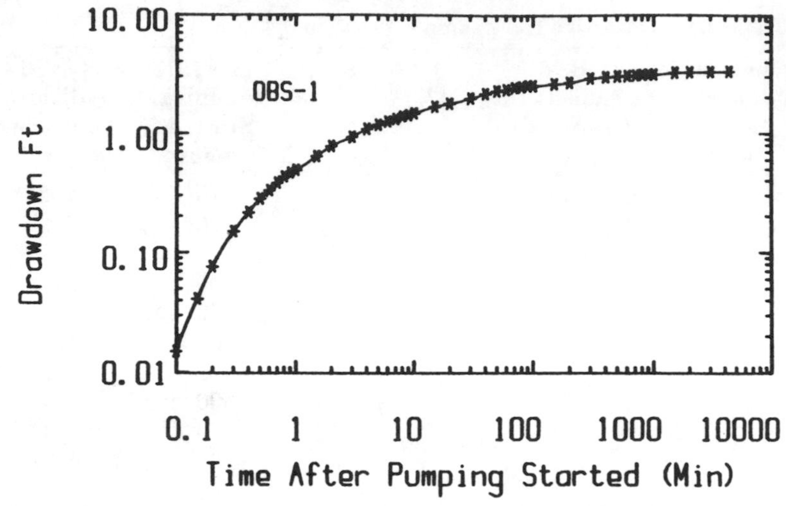

Figure 3.5. Time-drawdown graph for Example Problem 3.4.

tivity is moderately high, as is the aquitard hydraulic conductivity. The aquitard storativity (0.0004), based on the results of laboratory consolidation tests, is slightly higher than the storativity of the aquifer and the pretest conceptual modeling effort indicates aquitard storativity impacts are appreciable.

Time-drawdown data, adjusted for barometric pressure changes, for the two observation wells are presented in Table 3.5. Calculate aquifer transmissivity and storativity and aquitard vertical hydraulic conductivity.

A family of W(u,Gamma) versus 1/u type curves are presented in Figure F.5 of Appendix F. The time-drawdown curve for OBS-1 was found to be analogous to the Gamma = 0.008 type curve. Matchpoint coordinates and this value of Gamma were substituted into Equations 3.28–3.30. Match point coordinates are: W(u,Gamma) = 1.0, 1/u = 10, s = 1.1 ft, and t = 0.26 min. Calculated values of transmissivity, aquifer storativity, and aquitard vertical hydraulic conductivity are 104,180 gpd/ft, 0.0001, and 0.09 gpd/sq ft, respectively. The calculated value of transmissivity and Equation 1.1 with r_c = 0.2 ft indicates well storage capacity

Table 3.5. Database for Example Problem 3.5

Time after Pumping Started (min)	OBS-1 Adjusted Drawdown (ft)	OBS-2 Adjusted Drawdown (ft)
1	3.22	
2	3.89	
3	4.36	
4	4.58	
5	4.82	
6	4.97	
7	5.16	
8	5.35	
9	5.39	
10	5.42	
20	6.03	.18
30	6.41	.47
40	6.72	.77
50	6.96	1.05
60	7.03	1.38
70	7.14	1.66
80	7.20	1.88
90	7.28	2.17
100	7.35	2.32
200	7.91	3.96
300	8.06	4.72
400	8.24	5.30
500	8.52	5.61
600	8.67	6.02
700	8.76	6.30
800	8.82	6.53
900	8.90	6.72
1000	9.00	6.93
2000	9.38	7.81
2880	9.50	8.20

impacts were appreciable only during the first 2 minutes of the pumping test period. Aquitard leakage impacts are clear at the end of the test and become appreciable during the first few minutes of the pumping test period.

The adjusted drawdowns in aquifer OBS-1 and aquitard OBS-2 for a time period of 1000 minutes (9.0 and 6.9 ft,

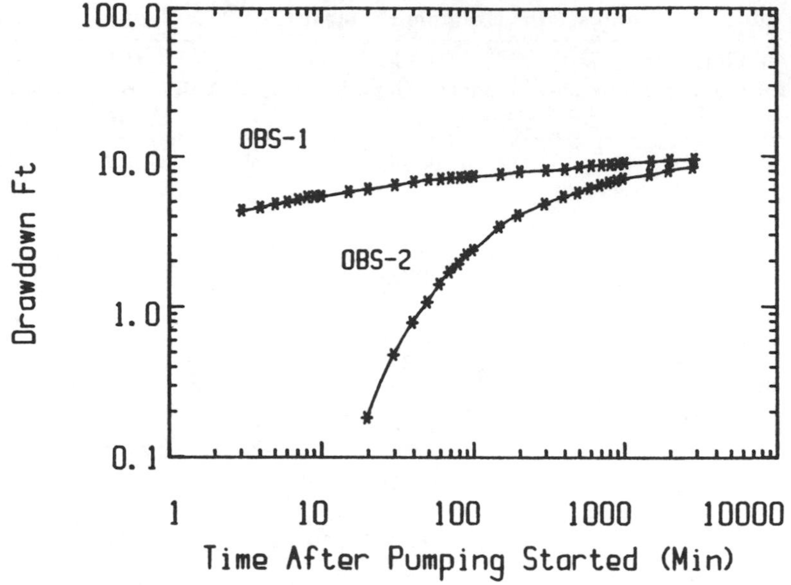

Figure 3.6. Time-drawdown graphs for Example Problem 3.5.

respectively), the calculated value of u (2.57×10^{-5}) (see Equation 2.7), calculated values of aquifer transmissivity and storativity, the calculated value of s_c/s (0.75), $1/u_c$ (25) from Table E.5 in Appendix E, and Equation 3.31 were utilized to determine the aquitard vertical hydraulic conductivity. This value of P_c is close to the value of P_c calculated from aquifer drawdown data and the problem solution is declared valid. Notice that drawdown in the aquitard observation well is negligible during the first 10 minutes of the pumping test period. Time-drawdown graphs are presented in Figure 3.6.

A 2-day pumping test was conducted with a production well and two observation wells in a water table aquifer which is 50 ft thick in Example Problem 3.6. There are no nearby aquifer boundaries or interfering production wells, the degree of aquifer stratification is high, and the hydraulic conductivity of the aquifer is high. The production well had a 0.5-ft radius and was open between aquifer depths of 30 and 50 ft. Observation well OBS-1 is located 35 ft from the production well and is open between aquifer depths of 10 and 12 ft. Observation well OBS-2 is located 400 ft from the production well and is open between aquifer depths of 30 and 50 ft. The production well constant discharge rate is 1000 gpm. Time-drawdown data for OBS-2, adjusted for dewatering impacts, and time-drawdown data for OBS-1, adjusted for dewatering impacts and partial penetration impacts with microcomputer program PT2, are presented in Table 3.6. Calculate aquifer transmissivity, specific yield, and the ratio P_V/P_H.

Families of $W(u_A,Beta)$ versus $1/u_A$ and $W(u_B,Beta)$ versus $1/u_B$ type curves are presented in Figures F.6 and F.7 of Appendix F. The time-drawdown curve for OBS-2 was found to be analogous to the Beta = 2.5 type curve during early, intermediate, and late time periods. Match point coordinates and the value of Beta were substituted into Equations 3.32–3.35. Match point coordinates are: $W(u_A,Beta) = 1.0$, $1/u_A = 10$, s = 0.45 ft, and t = 7 min; $W(u_B,Beta) = 1.0$, $1/u_B = 10$, s = 0.45 ft, and t = 1800 min. Calculated values of transmissivity, storativity, specific yield, and ratio P_V/P_H are 254,667 gpd/ft, 0.0004, 0.1, and $^1/_{25}$, respectively. The calculated value of storativity, based on limited early time-drawdown data, is weak. These values and Equations 1.1–1.3 with $r_c = 0.2$ ft indicate well storage capacity impacts are negligible after a time period of $^1/_2$ minute, partial penetration impacts are appreciable to a distance of 375 ft from the production well, and delayed gravity yield impacts are appreciable during the first 1330 minutes of the pumping test. Time-drawdown data for OBS-1 were adjusted for partial penetration impacts with transmissivity, storativity, specific yield, and ratio P_V/P_H values calculated with data for OBS-2 instead of constructing a family of type curves including partial penetration for OBS-1. For example, the partial penetration impact at the end of the test is -1.04 ft and the observed drawdown is 2.09 ft. Thus, the drawdown with full penetration is 3.13 ft. The adjusted time-drawdown curve for OBS-1 was found to be analogous to the Beta = 0.02 type curve. Match point

Table 3.6. Database for Example Problem 3.6

Time After Pumping Started (min)	OBS-1 Adjusted Drawdown (ft)	OBS-2 Adjusted Drawdown (ft)
0.5	1.25	0.02
0.75	1.31	0.03
1	1.35	0.03
2	1.41	0.04
3	1.42	0.05
4	1.42	0.05
5	1.42	0.05
6	1.42	0.05
7	1.44	0.05
8	1.46	0.05
9	1.48	0.05
10	1.50	0.05
20	1.53	0.06
30	1.57	0.06
40	1.65	0.07
50	1.72	0.08
60	1.76	0.09
70	1.80	0.09
80	1.82	0.10
90	1.84	0.10
100	1.86	0.11
200	2.12	0.18
300	2.33	0.24
400	2.46	0.30
500	2.56	0.35
600	2.62	0.42
700	2.73	0.47
800	2.82	0.53
900	2.88	0.56
1000	2.94	0.61
2160	3.13	0.93

coordinates and the value of Beta were substituted into Equations 3.32–3.35 to calculate values of transmissivity, storativity, specific yield, and ratio P_V/P_H, respectively. Match point coordinates are: $W(u_A,Beta) = 1.0$, $1/u_A = 100$, $s = 0.45$, and $t = 0.53$ min; $W(u_B,Beta) = 1.0$, $1/u_B = 100$,

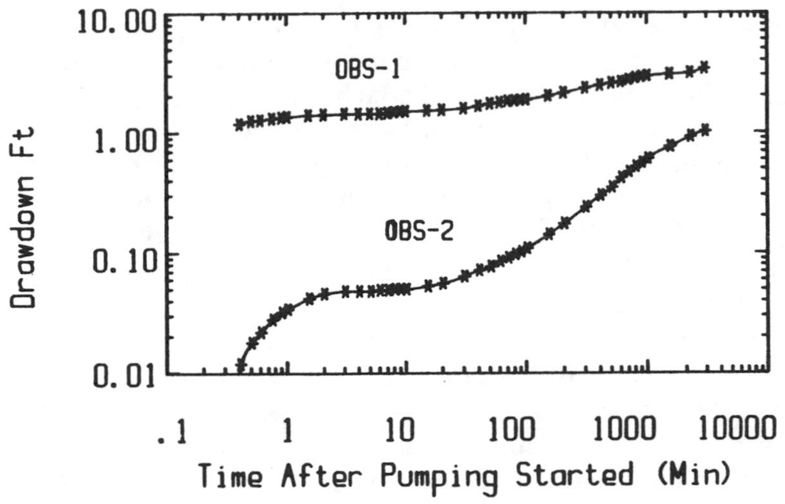

Figure 3.7. Time-drawdown graphs for Example Problem 3.6.

s = 0.45 ft, and t = 130 min. These values agree with the values calculated with data for OBS-2.

Adjusted drawdowns at the end of the test in OBS-1 and OBS-2 were plotted against values of r^2. This distance-drawdown curve was matched to the W(u) versus u type curve in Figure F.2 of Appendix F and match point coordinates and Equations 3.20 and 3.22 were used to calculate values of transmissivity and specific yield. These values agree with values determined with time-drawdown data and the problem solution is declared valid. Time-drawdown graphs are presented in Figure 3.7.

STRAIGHT LINE MATCHING TECHNIQUE

The straight line matching technique is based on modified model 1 equations (Cooper and Jacob, 1946, pp. 526–534) and the fact that graphs of W(u) versus the logarithm of time and W(u) versus the logarithm of 1/u describe straight lines when $u \leq 0.02$. The technique may be applied in cases where aquifer system conditions are similar to model 1 assumptions and well storage capacity, partial penetration, and delayed gravity yield impacts are negligible. The technique is particularly applicable to data for a production well and data collected under induced streambed infiltration conditions because u becomes small usually after a few minutes of pumping in the case of a production well and steady state conditions prevail after a relatively short pumping period in the case of recharge from a nearby stream.

Values of drawdown may be plotted against the logarithms of time after pumping started. A straight line may be fitted to portions of the time-drawdown semilogarithmic graph where $u \leq 0.02$ utilizing microcomputer program PT6. That program is based in part on a program developed by Poole et al. (1981, pp. 151-153) and utilizes the method of least squares. The interactive input section of the program prompts the user to enter the time-drawdown data selected for analysis in the form of x,y (time, drawdown) coordinates for points. The program finds the best-fit straight line to the data points and calculates the slope and zero-drawdown intercept of the line. Aquifer transmissivity and storativity are then calculated with slope and zero-intercept equations.

Aquifer storativity cannot be determined with any reasonable degree of accuracy from data for the production well because the effective radius of the production well is seldom known and drawdowns in the production well are often affected by well losses which cannot be determined precisely.

Values of drawdown for a specified common pumping test period in two or more observation wells at different distances from the production well may be plotted against the logarithms of the respective distances. A straight line may be fitted to portions of the distance-drawdown graph where $u \leq 0.02$ utilizing microcomputer program PT6. Aquifer transmissivity and storativity are calculated with slope and zero-drawdown intercept equations.

Scattered drawdown data are sometimes interpreted as

describing a straight line when actually they plot as a gentle curve. After tentative values of transmissivity and storativity have been calculated, the segment of the data where $u \leq 0.02$ should be determined and compared with the segment of data through which the straight line was drawn. The time that must elapse before the straight line matching technique can be properly applied to pumping test data is as follows (Walton, 1962, p. 9):

$$t_{s1} = 1.35 \times 10^5 r^2 S/T \tag{3.38}$$

where t_{s1} = time that must elapse before $u \leq 0.02$, in min
r = distance from production well, in ft
S = aquifer storativity, dimensionless
T = aquifer transmissivity, in gpd/ft

The time that must elapse before a semilogarithmic time-drawdown or distance-drawdown graph will yield a straight line may vary from several minutes under artesian conditions to more than one day under water table conditions.

Calculation of aquifer storativity by the straight line matching method may involve appreciable error. The zero-drawdown intercept is poorly defined where the slope is small. Intercepts often occur at points where the values of time are very small and minor deviations in extrapolating the straight line will result in large variations in calculated values of storativity.

Slope and Zero-Drawdown Intercept Equations

Time-drawdown slope and zero-drawdown intercept equations are (see Cooper and Jacob, 1946, pp. 526–534):

$$T = 264Q/s_1 \tag{3.39}$$

$$S = Tt_0/(4790r^2) \tag{3.40}$$

where T = aquifer transmissivity, in gpd/ft
Q = discharge rate, in gpm

s_1 = drawdown difference per log cycle, in ft
S = aquifer storativity, dimensionless
t_0 = zero-drawdown intercept, in min
r = distance from production well, in ft

Distance-drawdown slope and zero-drawdown intercept equations are (see Cooper and Jacob, 1946, pp. 526–534):

$$T = 528Q/s_1 \tag{3.41}$$

$$S = Tt/(4790r_0^2) \tag{3.42}$$

where t = time after pumping started, in min
r_0 = zero-drawdown intercept, in ft

Induced Streambed Infiltration

When a production well is pumped near a stream hydraulically connected to an aquifer, water is first withdrawn from storage within the aquifer in the immediate vicinity of the production well. The cone of depression then spreads, drawing water from storage within an increasing area of influence. Water levels in the vicinity of the stream are lowered, and more and more of the water which under natural conditions would have discharged into the stream as groundwater runoff is diverted toward the production well. Water levels are lowered below the surface of the stream in the immediate vicinity of the production well, and the aquifer is recharged by the influent seepage of surface water (see Walton, 1963).

The cone of depression continues to grow until it intercepts sufficient area of the streambed and is deep enough so that induced streambed infiltration balances discharge. If the hydraulic conductivity of the streambed is high, the cone of depression may extend only partway across the stream; if the hydraulic conductivity of the streambed is low, the cone of depression may expand across and beyond the stream.

Recharge by induced streambed infiltration takes place

over an area of the streambed. However, to make flow problems amenable to mathematical treatment, the area is replaced by a recharging image well. The assumption is made that water levels in the aquifer will behave the same way whether recharge occurs over an area or through a recharging image well (see previous discussion on image well theory) located at an effective distance from the production well. It is further assumed that streambed partial penetration and aquifer stratification impacts are integrated into that effective distance. Representative values of induced streambed infiltration are presented in Table C.6 of Appendix C.

Aquifer transmissivity is determined with steady-state distance-drawdown data at the end of the pumping test for observation wells on a line through and close to the production well and parallel to the stream. These observation wells should be approximately equidistant from the recharging image well, and the impacts of the induced streambed infiltration on water levels in these wells should be equal. Thus, the hydraulic gradient toward the production well along the line is not distorted to any appreciable degree and closely describes the hydraulic gradient that would exist if the aquifer were infinite in areal extent. A plot of drawdown in the observation wells parallel to the stream versus the logarithm of the distances from the production well will yield a straight line graph. The slope of the straight line and the discharge rate are inserted into Equation 3.41 and aquifer transmissivity is calculated.

Although the hydraulic gradient is not distorted, the total values of drawdown in the observation wells are much less than they would be without recharge from the stream. Aquifer storativity (specific yield) cannot be determined directly from the distance-drawdown graph. Instead, specific yield is determined by the method of successive approximations based on drawdown data at the end of the test and the location of the recharging image well.

The distance from the production well to the recharging

image well is calculated with the following equation (see Rorabaugh, 1956, pp. 101–169):

$$D_i = 2\{[r^2(10^{Ts/(528Q)})^2 - r^2]/4\}^{1/2} \qquad (3.43)$$

where D_i = distance from production well to recharging image well, in ft
r = distance from the production well to observation well, in ft
T = aquifer transmissivity, in gpd/ft
s = drawdown in observation well, in ft
Q = production well discharge, in ft

Equation 3.43 is valid for observation wells on a line parallel to the stream and the distance D_i is measured at a right angle to the stream (see previous discussion concerning the image well theory).

Microcomputer Programs

With the recharging image well located and aquifer transmissivity calculated, aquifer specific yield may be determined using observation well data at the end of the test and Equations 3.1, 3.2, and 3.13–3.16. The production, observation, and image wells are drawn to scale on a map and the distances between the image well and the observation wells are determined. Several values of aquifer specific yield are assumed and microcomputer program PT7 is utilized to determine observation well drawdowns for each assumed value. Calculated drawdown values are compared with actual drawdown values, and that specific yield used to calculate drawdowns equal to actual drawdowns is assigned to the aquifer. The interactive input section of microcomputer program PT7 prompts the user to enter values of production well discharge which are equal to the image well recharge rate, aquifer transmissivity, aquifer storativity, distance between the observation well and the production well, distance between the image well and the observation well, and the time after pumping started. Total drawdowns in observation wells due to both production and image well impacts are generated and displayed as output by the program.

The rate of stream depletion or the rate of recharge by induced streambed infiltration is calculated with microcomputer program PT8, which is based on equations presented by Jenkins (1968, p. 16). The interactive input section of the program prompts the user to enter values of aquifer transmissivity, aquifer storativity, time after pumping started, production well discharge rate, and distance between the production and recharging image wells. The rate of stream depletion is generated and displayed as output by the program.

The production well, recharging image well, and streambed are drawn to scale on a map. A grid is superposed over the map. Several points within the streambed up and down stream are selected for calculation of drawdown beneath the streambed. Values of drawdown at the points are then determined with microcomputer program PT9. The interactive input section of that program prompts the user to enter values of aquifer transmissivity, specific yield, production well discharge rate, time after pumping started, x,y coordinates of the production and image wells, and grid spacing. Total drawdowns due to the production and image wells are generated at grid line intersections by the program and displayed as output. Values of drawdown at the selected points within the streambed may be interpolated from grid intersection values.

The reach of the streambed (L_r) within the influence of the production well is ascertained by noting the points up and down stream where drawdown beneath the streambed is negligible (<0.01 ft). The area of induced streambed infiltration is then equal to the product of L_r and the average distance between the shoreline and the recharge boundary or the average width of the streambed depending upon the position of the recharge boundary.

The induced streambed infiltration rate is calculated with the following equation (Walton, 1963):

$$I_a = 6.3 \times 10^7 Q_r/A_r \qquad (3.44)$$

where I_a = average induced streambed infiltration rate, in gpd/acre
Q_r = rate of stream depletion, in gpm
A = area of induced streambed infiltration, in sq ft

The average head loss due to the vertical percolation of water through the streambed may be determined from data for observation wells installed within the streambed area of induced infiltration at depths just below the streambed. In many cases, the installation of observation wells in the stream channel is impractical and the average head loss must be estimated with Equations 3.1–3.2 and 3.13–3.16 and microcomputer program PT9. Many points within the reach of the streambed influenced by the production well are located on a map and drawdowns at the points are then interpolated from grid intersection values of drawdown and averaged.

The average induced streambed infiltration rate per unit area per foot of head loss may be estimated with the following equation (Walton, 1963):

$$I_h = I_a/h_r \qquad (3.45)$$

where I_h = average induced streambed infiltration rate, in gpd/acre/ft
h_r = average head loss within the streambed area of induced infiltration, in ft

The induced streambed infiltration rate per foot of head loss varies with the temperature of the surface water. A decline in the temperature of surface water of 1°F will decrease the rate about 1.5% (Rorabaugh, 1956, pp. 101–169) through the range generally encountered in practical problems. The induced streambed infiltration rate for any particular surface water temperature may be calculated with values of dynamic viscosity presented in Table E.7 of Appendix E and the following equation (Walton, 1963):

$$I_t = I_h V_a / V_t \tag{3.46}$$

where I_t = average induced streambed infiltration rate for a particular surface water temperature, in gpd/acre/ft
V_a = dynamic viscosity at temperature of surface water during pumping test, in poise-second units
V_t = dynamic viscosity at a particular temperature of surface water, in poise-second units

Example Problems

An 8-hour pumping test was conducted with a production well and one observation well in a nonleaky artesian aquifer system in Example Problem 3.7. There are no nearby aquifer boundaries or interfering production wells and the wells fully penetrate the aquifer. Observation well OBS-1 is located 234 ft from the production well and the production well has a radius of 0.33 foot. The constant production well discharge rate is 50 gpm. Time-drawdown data for the production and observation wells are presented in Table 3.7. Calculate aquifer transmissivity and storativ-

Table 3.7. Database for Example Problem 3.7

Time After Pumping Started (min)	Production Well Drawdown (ft)	OBS-1 Drawdown (ft)
10	2.88	0.46
20	3.01	0.59
40	3.14	0.71
70	3.24	0.82
150	3.38	0.96
300	3.51	1.09
480	3.60	1.17

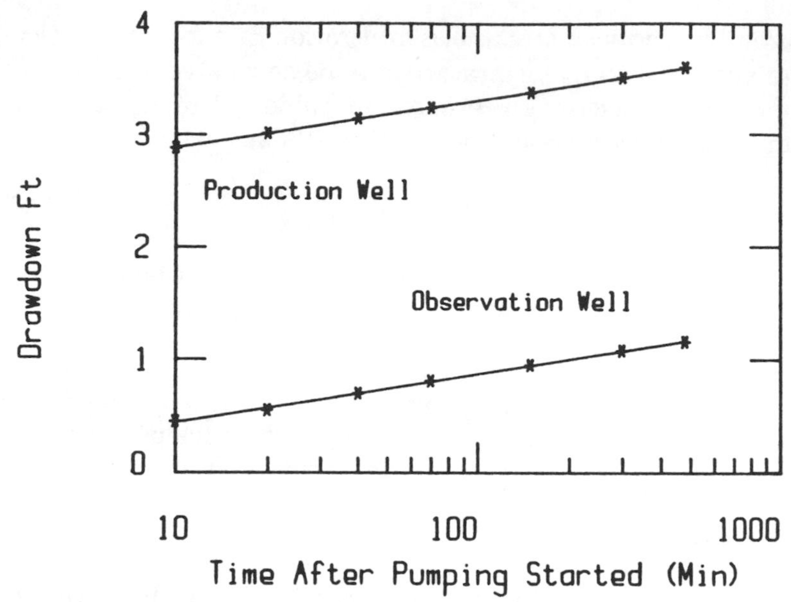

Figure 3.8. Time-drawdown graphs for Example Problem 3.7.

ity using the straight line matching method and microcomputer program PT6.

The slopes of the straight lines through both time-drawdown graphs are 0.42 ft; the zero-drawdown intercept for OBS-1 is 0.824 minute. Values of aquifer transmissivity and storativity calculated with Equations 3.39 and 3.40 are 31,429 gpd/ft and 1×10^{-4}, respectively. Aquifer storativity cannot be calculated from production well data because the effective radius and well loss coefficient for the production well are unknown. According to Equation 3.38, u was ≤ 0.02 before the first water level measurement was made in the production well and 24 minutes after pumping started in OBS-1. The time-drawdown graphs for the two wells are presented in Figure 3.8.

A 3-day pumping test was conducted with a production well and three observation wells in a water table aquifer system in Example Problem 3.8. The production and observation wells fully penetrate the aquifer. The production

Table 3.8. Database for Example Problem 3.8

Distance from Production Well to Observation Well (ft)	Drawdown (ft)
50	.81
100	.54
250	.23

well constant discharge rate was 100 gpm. The thickness of the aquifer is 50 ft and the radius of the production well was 0.33 ft. A lake hydraulically connected to the aquifer system (source of recharge) is located 100 ft from the production well. All observation wells are on a line through the production well and parallel to the lake shore. Distances from the production well to OBS-1, OBS-2, and OBS-3 are 50, 100, and 200 ft, respectively. The temperature of the surface water is 54°F. Distance-drawdown data for the end of the test are presented in Table 3.8. Calculate the aquifer transmissivity and specific yield and the lakebed induced infiltration rate.

The slope of the straight line through the distance-drawdown graph is 0.90 ft and the aquifer transmissivity calculated with Equation 3.41 is 58,667 gpd/ft. The aquifer specific yield calculated with microcomputer program PT7 is 0.05. Lake depletion at the end of the test was calculated to be 84 gpm with microcomputer program PT8. The distance from the production well to the recharging image well simulating lake induced infiltration was calculated with Equation 3.43 to be 400 ft. Thus, the effective line of recharge is 200 ft from the production well or 100 ft offshore. Drawdowns between the lake shore and the effective line of recharge ranged from 0.42 to 0.01 ft and average 0.05 ft within the area of influence of pumping according to microcomputer program PT9. Drawdown was appreciable 1000 ft up and down the lake shore. The area of lakebed induced infiltration was 200,000 sq ft. The lakebed induced

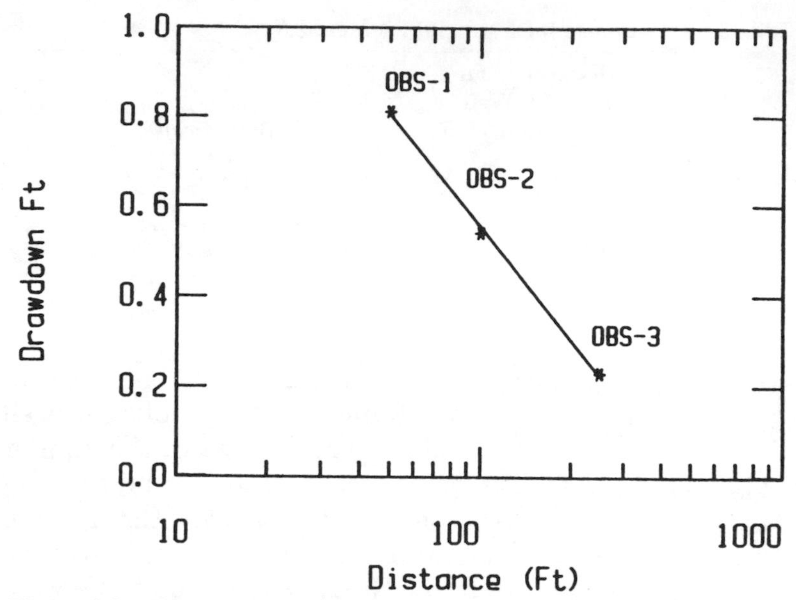

Figure 3.9. Distance-drawdown graph for Example Problem 3.8.

infiltration rate is 26,460 gpd/acre based on Equation 3.44 and the lakebed induced infiltration rate per foot of head loss beneath the lakebed is 529,200 gpd/acre/ft based on Equation 3.45. The lakebed induced infiltration rate per foot of head loss would be 763,033 gpd/acre/ft if the temperature of the surface water were 79°F (see Equation 3.46). Calculations made with Equations 1.1 and 1.3 indicate that well storage impacts were negligible with $r_c = 0.1$ ft after 1 minute of pumping and delayed gravity yield impacts were negligible with $P_V/P_H = 1/10$ before the end of the test. The distance-drawdown graph is presented in Figure 3.9.

4

Step Drawdown Test Analysis

One of the components of drawdown in a production well is well loss due to the turbulent flow of water through the well face and inside the casing to the pump bowls. Well loss is estimated using data collected during a step drawdown test which consists of operating the pump in the production well during three successive periods usually 1 hour in duration at constant fractions of full capacity. Water levels in the production well are measured at frequent intervals as is the discharge rate.

ANALYTICAL MODEL

The analytical well loss model assumes well loss is proportional to the square of the production well discharge rate and the production well is stable. Well loss may be represented approximately by the following equation (Jacob, 1946, pp. 1047-1070):

$$s_w = CQ^2 \tag{4.1}$$

where s_w = drawdown component due to well loss, in ft
C = well loss coefficient, in sec^2/ft^5
Q = production well discharge rate, in cfs (1 cfs = 449 gpm)

The well loss coefficient may be estimated with the following equation (Jacob, 1946, pp. 1047–1070):

$$C = (s_n/Q_n - s_{n-1}/Q_{n-1})/(Q_{n-1} + Q_n)$$

where s_n = increment of drawdown for pumping rate number n, in ft
Q_n = increase in discharge for pumping rate number n, in cfs
s_{n-1} = increment of drawdown for pumping rate number n−1, in ft
Q_{n-1} = increase in discharge for pumping rate number n−1, in cfs
n = pumping rate number (1, 2, or 3)

Increments of drawdown are determined by taking in each case the difference between the observed water level in the production well and the extrapolation of the preceding water level curve (antecedent trend). s_n and s_{n-1} values must be for the same time interval.

For steps 1 and 2:

$$C_{1 \text{ and } 2} = (s_2/Q_2 - s_1/Q_1)/(Q_1 + Q_2) \quad (4.2)$$

For steps 2 and 3:

$$C_{2 \text{ and } 3} = (s_3/Q_3 - s_2/Q_2)/(Q_2 + Q_3) \quad (4.3)$$

Values of C for pumping test production wells are generally less than 10 sec^2/ft^5 and are often about 2.0 sec^2/ft^5. Calculation of the well loss coefficient may be impractical with discharge rates of a few gallons per minute or less.

Well Loss Microcomputer Program

Well loss coefficients and well loss may be calculated with microcomputer program PT10. The interactive input section of the program prompts the user to enter three values of discharge rate and associated changes in drawdown and a value of discharge for which well loss is to be calculated. The program generates and displays as

Table 4.1. Database for Example Problem 4.1

Discharge Rate (gpm)	Drawdown (ft)
100	3.35
151	5.14
199	6.86

output values of well loss coefficient for steps 1 and 2, 2 and 3, 1+2 and 3, and 2+3 and 1. In addition, the program calculates the well loss for the specified discharge rate.

Well loss coefficients are calculated for steps 1+2 and 3 and 2+3 and 1 because sometimes production wells are unstable and clogging or development occurs between discharge rate changes. When this happens, the value of well loss coefficient calculated for steps 1 and 2 may be much greater or less than that calculated for steps 2 and 3. Solution of Equation 4.3 may be impossible; however, the well loss coefficient may be calculated with data for steps 1+2 and 3 or 2+3 and 1.

Example Problem

A step drawdown test was conducted with a production well in stable condition in Example Problem 4.1. Measured discharge rates and associated drawdowns for three steps are presented in Table 4.1. Calculate the well loss coefficient and well loss for the 151 gpm discharge rate utilizing microcomputer program PT10.

The average calculated value of well loss coefficient is 2.0 sec^2/ft^5 and well loss at the 151 gpm discharge rate based on Equation 4.1 is 0.23 ft.

5

Case Studies

Case studies presented herein were selected to provide further insight into the practical application of techniques for pumping test analysis under a variety of aquifer system conditions. Each pumping test discussed has its own unique characteristics; however, commonalities of logic do thread their way through all case studies. Nonleaky artesian, leaky artesian, water table, and induced streambed infiltration conditions are covered.

Experience has shown that all data are equally important in pumping test analysis. Segments of data should be rejected or filtered out only with due justification such as bad water level records due to a stuck float in an observation well. Early- and late-time drawdown data are equally important; type curves should not be automatically matched to either early or late data without due reason.

NONLEAKY ARTESIAN AQUIFER SYSTEM

A pumping test (see Walton, 1962, pp. 33–34) was conducted on July 2, 1953 at Gridley, Illinois, in Case Study 5.1. A group of fully penetrating village wells was used. In general, 270 ft of clay overlie the 18-ft thick fine sand and some gravel aquifer which is underlain by shale. There are no nearby interfering production wells or aquifer system boundaries. The effects of pumping well 3 were measured in observation wells 1 and 2. Observation well 1 was located 824 ft west of production well 3 and observation well 2 was located 850 ft west of production well 3. Pumping was started at 9:45 a.m. on July 2 and was continued at a constant rate of 220 gpm for about 8 hours until 6:02 p.m. Time-drawdown data for the production well and observation well 1 adjusted for minor barometric pressure changes are presented in Table 5.1.

A straight line was fitted to a semilogarithmic time-drawdown graph for production well 3. The slope of the straight line was used to calculate aquifer transmissivity. Aquifer storativity cannot be calculated with the graph because the effective radius of the

Table 5.1. Database for Case Study 5.1

Time After Pumping Started (min)	Well 1 Adjusted Drawdown (ft)	Time After Pumping Started (min)	Well 3 Adjusted Drawdown (ft)
3	0.3	15	24.8
5	0.7	25	25.5
8	1.3	45	26.6
12	2.1	60	27.3
20	3.2	76	28.0
24	3.6	90	28.2
30	4.1	132	29.0
38	4.7	166	29.5
47	5.1	195	30.3
50	5.3	256	30.5
60	5.7	282	30.6
70	6.1	314	30.7
80	6.3	360	30.8
90	6.7	430	31.5
100	7.0		
130	7.5		
160	8.3		
200	8.5		
260	9.2		
320	9.7		
380	10.2		
500	10.9		

production well is unknown and total drawdown in the production well is affected by well loss.

The type curve for model 1 (Figure F.1 in Appendix F) was matched to the logarithmic time-drawdown graph for observation well 1. Match point coordinates were used to calculate aquifer transmissivity and storativity.

Calculations for the production well and observation well 1 are as follows:

Production Well

$$Q = 220 \text{ gpm}, \qquad s_1 = 4.8 \text{ ft}$$

$$T = 2.64 \times 10^2(2.2 \times 10^2)/4.8 \qquad T = 12{,}100 \text{ gpd/ft}$$

Observation Well 1

$$Q = 220 \text{ gpm}, \qquad r = 824 \text{ ft}, \qquad m = 18 \text{ ft}$$

$$\text{Match Pt. Coord. } W(u) = 0.1, \qquad 1/u = 1.0$$
$$s = 0.25 \text{ ft}, \qquad t = 4.3 \text{ min}$$

$$T = 1.146 \times 10^2(2.2 \times 10^2)1 \times 10^{-1}/2.5 \times 10^{-1}$$
$$T = 10{,}100 \text{ gpd/ft}$$

$$S = 1.01 \times 10^4(1.0)4.3/[2.693 \times 10^3(6.8 \times 10^5)]$$
$$S = 0.00002$$

Time-drawdown graphs for the production well and observation well 1 are presented in Figures 5.1 and 5.2.

The average calculated aquifer transmissivity and storativity are 11,000 gpd/ft and 2.2×10^{-5}, respectively. Well storage capacity impacts are appreciable during the first 5 minutes of pumping; in analyzing test data, emphasis was placed on data after this time. The pumping test sampled a cylindrical volume of the aquifer with a height of 18 ft (aquifer thickness) and a radius of about 22,000 ft. Aquifer hydraulic conductivity is 611 gpd/sq ft, which is indicative of a fine sand aquifer. (See Table C.1 in Appendix C.)

A pumping test was conducted in Case Study 5.2 on June

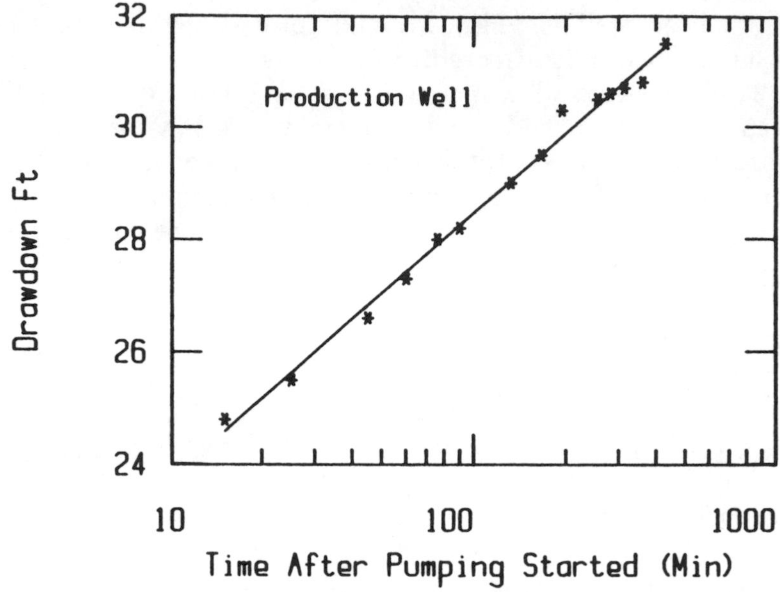

Figure 5.1. Time-drawdown graph for production well in Case Study 5.1.

2, 1960, using a group of fully penetrating wells located about 1.5 miles north of the corporate limits of the village of St. David, Illinois (see Walton, 1962, pp. 47–48). In general, 40 ft of sandy clay overlie an 8-ft-thick fine sand and some gravel aquifer which is underlain by shale. The aquifer pinches out (barrier boundaries) east and west of the production well. There are no nearby interfering production wells. The effects of discharging the production well were measured in observation wells 1 and 2. Observation well 1 was located 165 ft east of the production well and observation well 2 was located 20 ft west of the production well. Pumping was started at 10:00 a.m. and was continued for six hours at a constant rate of 62 gpm. Time-drawdown data for the observation wells adjusted for minor changes in barometric pressure are presented in Table 5.2.

The type curve for model 1 (Figure F.1 in Appendix F) was matched to early portions of the logarithmic time-

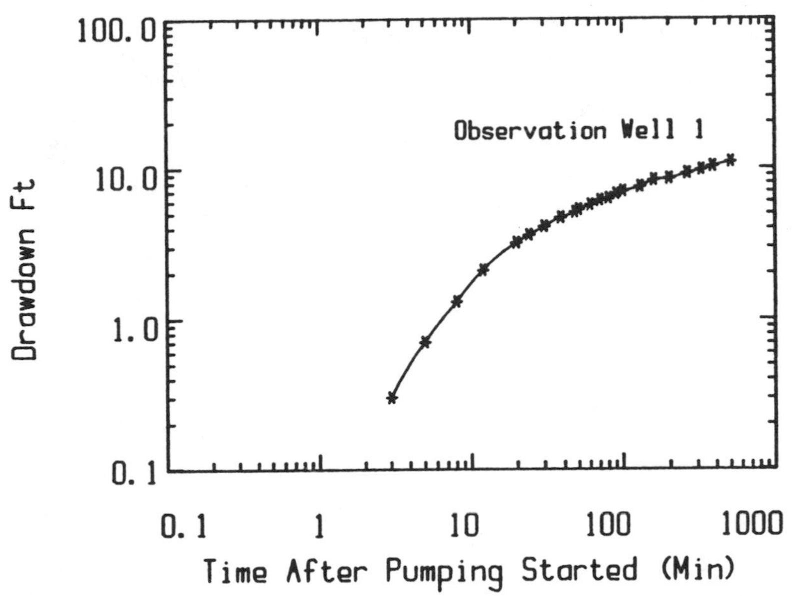

Figure 5.2. Time-drawdown graph for observation well 1 in Case Study 5.1.

drawdown graphs for observation wells 1 and 2 before boundary impacts became appreciable. Aquifer transmissivity and storativity were calculated with match point coordinates. After about 12 minutes of pumping, the time-rate of drawdown in the observation wells increased and data deviated upward from type curve traces, indicating the presence of a barrier boundary. The model 1 type curve (Figure F.1 in Appendix F) was matched to later time-drawdown data affected by the barrier boundary. The divergences of the first and second type curve traces were used to determine the distances from the observation wells to the first barrier boundary image well.

After about 70 to 100 minutes of pumping, the time-rate of drawdown in the observation wells again increased, and data deviated upward from the second type curve trace, indicating the presence of a second barrier boundary. The model 1 type curve was matched to late time-drawdown data affected by both barrier boundaries. The divergences

Table 5.2. Database for Case Study 5.2

Time After Pumping Started (min)	Well 1 Adjusted Drawdown (ft)	Time After Pumping Started (min)	Well 2 Adjusted Drawdown (ft)
5	1.00	2	4.60
6	1.19	3	4.97
7	1.25	4	5.32
8	1.43	5	5.65
9	1.58	6	5.91
10	1.72	7	6.08
12	1.94	8	6.20
15	2.32	9	6.33
18	2.64	10	6.50
20	2.85	15	7.62
25	3.26	20	8.21
30	3.61	25	8.75
40	4.42	30	9.23
50	5.00	40	10.02
60	5.51	50	10.55
71	6.12	60	11.25
80	6.50	70	11.74
90	6.91	80	12.21
100	7.28	90	12.54
152	8.91	100	12.80
195	9.90	150	14.35
254	11.15	195	15.38
300	11.64	255	16.57
360	12.60	300	17.14
		360	17.78

of the second and third type curve traces were used to determine the distances from the observation wells to the second barrier boundary image well.

The distances from each observation well to the image wells were scribed as arcs from the respective observation wells. Theoretically, the arcs should intersect at common points, but the real aquifer is not a vertically bounded aquifer (as assumed in the image well theory), and as a result the arcs and their intersections are dispersed. In addition, only two observation wells are available, so that the exact locations of the image wells, and therefore the barrier boundaries, cannot be determined with test data alone.

Logs of wells indicate that the aquifer occurs as a fill in a buried valley in shale bedrock. Based on pumping test and geologic data, the barrier boundaries were located at positions east and west of the production well. The boundaries represent a rectangular section which is equivalent hydraulically to the real aquifer system.

Calculations for the observation wells are as follows:

Observation Well 1

$$Q = 62 \text{ gpm}, \quad r = 165 \text{ ft}, \quad m = 8 \text{ ft}$$

$$\text{Match Pt. Coord. } W(u) = 1.0, \quad 1/u = 1.0, \quad s = 1.45 \text{ ft}$$
$$t = 2 \text{ min}$$

$$T = 1.146 \times 10^2(62)1.0/1.45 \qquad T = 4900 \text{ gpd/ft}$$

$$S = 4.9 \times 10^3(1.0)2/[2.693 \times 10^3(165)^2] \qquad S = 0.000134$$

$$s_{i1} = 1.0 \text{ ft}, \qquad t_{i1} = 44 \text{ min}$$

$$W(u_{i1}) = 4.9 \times 10^3(1.0)/[1.146 \times 10^2(62)]$$
$$W(u_{i1}) = 0.6896$$

$$u_{i1} = 4.12 \times 10^{-1}$$

$$r_{i1} = \{[4.9 \times 10^3(44)4.12 \times 10^{-1}]/[2.693 \times 10^3(1.34 \times 10^{-4})]\}^{1/2}$$

$$r_{i1} = 496 \text{ ft}$$

$$s_{i2} = 1.0 \text{ ft}, \qquad t_{i2} = 154 \text{ min}$$

$$W(u_{i2}) = 0.6896, \qquad u_{i2} = 4.12 \times 10^{-1}$$

$$r_{i2} = \{[4.9 \times 10^3(154)4.12 \times 10^{-1}]/[2.693 \times 10^3(1.34 \times 10^{-4})]\}^{1/2}$$

$$r_{i2} = 928 \text{ ft}$$

Observation Well 2

$$Q = 62 \text{ gpm}, \quad r = 20 \text{ ft}, \quad m = 8 \text{ ft}$$

$$\text{Match Pt. Coord. } W(u) = 1.0, \quad 1/u = 100, \quad s = 1.45 \text{ ft}$$
$$t = 5.6 \text{ min}$$

$$T = 1.146 \times 10^2(62)1.0/1.45 \qquad T = 4900 \text{ gpd/ft}$$

$$S = 4.9 \times 10^3(0.01)5.6/[2.693 \times 10^3(20)^2] \qquad S = 0.00025$$

$$s_{i1} = 3.6 \text{ min}, \qquad t_{i1} = 180 \text{ min}$$

$$W(u_{i1}) = 4.9 \times 10^3(3.6)/[1.146 \times 10^2(62)]$$
$$W(u_{i1}) = 2.4827$$

$$u_{i1} = 4.92 \times 10^{-2}$$

$$r_{i1} = \{[4.9 \times 10^3(180)4.92 \times 10^{-2}]/[2.693 \times 10^3(2.5 \times 10^{-4})]\}^{1/2}$$

$$r_{i1} = 253 \text{ ft}$$

$$s_{i2} = 3.0 \text{ ft} \qquad t_{i2} = 1000 \text{ min}$$

$$W(u_{i2}) = 4.9 \times 10^3(3)/[1.146 \times 10^2(62)] \qquad W(u_{i2}) = 2.0689$$

$$u_{i2} = 7.72 \times 10^{-2}$$

$$r_{i2} = \{[4.9 \times 10^3(1.0 \times 10^3)7.72 \times 10^{-2}]/[2.693 \times 10^3(2.5 \times 10^{-4})]\}^{1/2}$$

$$r_{i2} = 750 \text{ ft}$$

Time-drawdown graphs for the observation wells are presented in Figure 5.3.

The average values of aquifer transmissivity and storativity are 4900 gpd/ft and 0.0002, respectively. Based on an average aquifer thickness of 8 ft, the hydraulic conductivity is 613 gpd/sq ft. Available data indicate that the aquifer is a thin strip of fine sand and some gravel approximately 600 ft wide which trends northeast to southwest through the pumping test area. Well storage capacity impacts were appreciable until about 11 minutes after pumping started.

A pumping test was conducted in Case Study 5.3 in November 1979 using a small-diameter production well located in southwestern Indiana (see Davis and Walton, 1982, pp. 841–848). The production well encountered 93 ft of unconsolidated deposits, 207 ft of sandstone, 30 ft of limestone, 126 ft of shale, and 4 ft of coal. The casing with a radius of 0.25 ft extended from land surface 97 ft through the unconsolidated deposits and into bedrock.

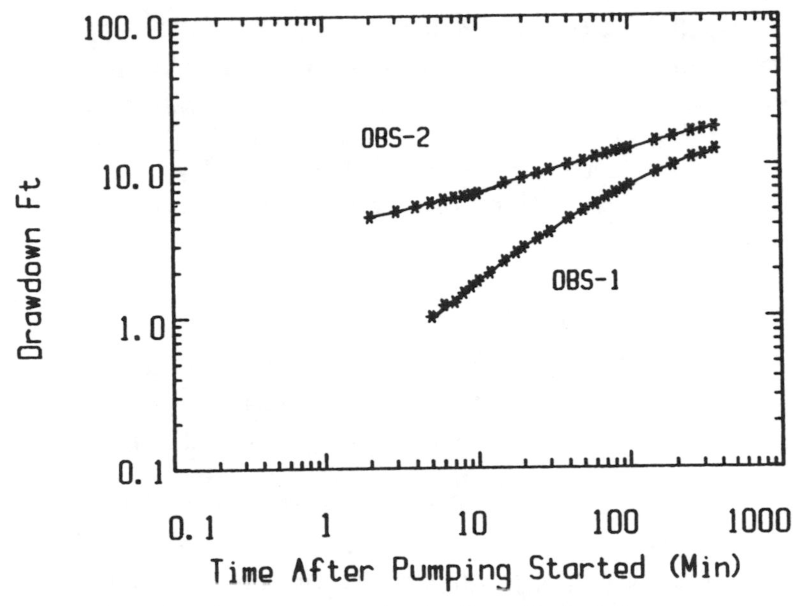

Figure 5.3. Time-drawdown graphs for Case Study 5.2.

The pump in the production well was discharged at a constant rate of 9 gpm for a period of 6 hours, and water levels in the production well were measured at frequent intervals. Recovery was also observed for a period of 6 hours. Time-recovery data for the production well are presented in Table 5.3.

Table 5.3. Database for Case Study 5.3

Time After Pumping Stopped (min)	Adjusted Recovery (ft)	Time After Pumping Stopped (min)	Adjusted Recovery (ft)
1	5.91	23	107.32
2	16.36	35	126.77
3	18.98	56	150.01
5	33.55	79	163.37
7	46.11	110	179.86
9	57.02	170	191.02
11	65.23	240	201.13
16	84.87	300	203.74

The Rho = 1 type curve for model 2 (Figure F.3 in Appendix F) was matched to the time-recovery logarithmic graph, and aquifer transmissivity and storativity were calculated to be 74 gpd/ft and 1.03×10^{-4}, respectively. Well storage capacity impacts were appreciable during the entire pumping period. Calculations for the production well are as follows:

$$Q = 9 \text{ gpm}, \qquad r_w = 0.219 \text{ ft}$$

$$\text{Match Pt. Coord. } W(u,S,Rho) = 1.0, \qquad 1/u = 10^5,$$
$$s = 14 \text{ ft}, \qquad t = 18 \text{ min}$$

$$T = 1.146 \times 10^2(9)1.0/14 \qquad T = 73.7 \text{ gpd/ft}$$

$$S = 73.7(18)/[2.693 \times 10^3(4.78 \times 10^{-2})10^5] \; S = 1.03 \times 10^{-4}$$

The time-recovery graph for the production well is presented in Figure 5.4.

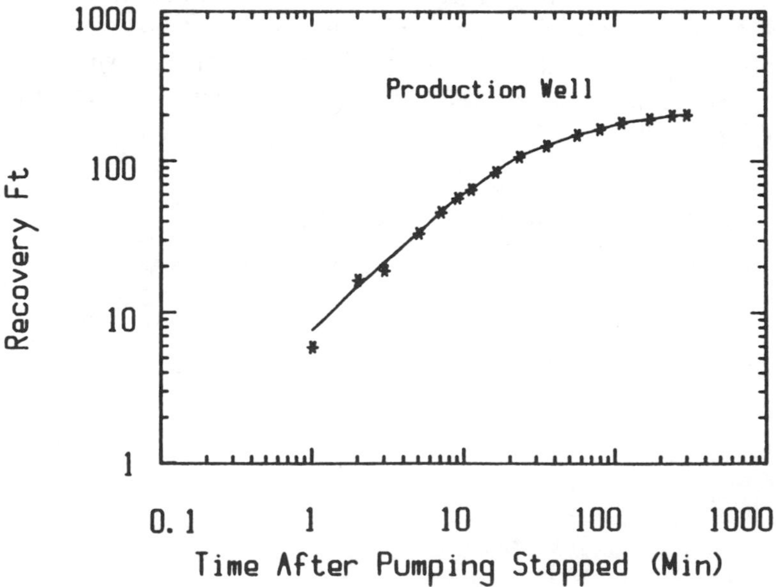

Figure 5.4. Time-recovery graph for Case Study 5.3.

Table 5.4. Database for Case Study 5.4

Time After Pumping Started (min)	Well 1 Adjusted Drawdown (ft)
5	0.76
28	3.30
41	3.59
60	4.08
75	4.39
244	5.47
493	5.96
669	6.11
958	6.27
1129	6.40
1185	6.42

LEAKY ARTESIAN AQUIFER SYSTEM

A pumping test was conducted in Case Study 5.4 on July 2 and 3, 1951, using a group of fully penetrating wells located about 1 mile southwest of the corporate limits of the village of Dietrich, Illinois (see Walton, 1962, pp. 31–33). In general, 22 ft of sandy clay overlie an 8-ft-thick fine sand and some gravel aquifer which is underlain by shale. The thickness of the aquitard is 14 ft. There are no nearby interfering production wells or aquifer boundaries. The effects of discharging a production well were measured in three observation wells. Observation well 1 was located 96 ft northwest of the production well, observation well 2 was located 92 ft southwest of the production well, and observation well 3 was located 245 ft northeast of the production well. Pumping was started at 2:10 p.m. on July 2 and was continued for a period of 20 hours at a constant rate of 25 gpm until 10:00 a.m. on July 3. Time-drawdown data for observation well 1 are presented in Table 5.4.

The time-drawdown curves for the observation wells were matched to appropriate model 3 type curves (Figure F.4 in Appendix F), and match point coordinates and the selected value of r/B were used to calculate aquifer transmissivity and storativity and aquitard vertical hydraulic conductivity. Calculations for observation well 1 are as follows:

$$Q = 25 \text{ gpm}, \quad r = 96 \text{ ft}, \quad m = 8 \text{ ft}, \quad m_c = 14 \text{ ft}$$
$$r/B = 0.22$$
$$\text{Match Pt. Coord. } W(u,r/B) = 1.0, \quad 1/u = 10, \quad s = 1.9 \text{ ft}$$
$$t = 33 \text{ min}$$
$$T = 1.146 \times 10^2(25)1.0/1.9 \qquad T = 1508 \text{ gpd/ft}$$
$$S = 1.51 \times 10^3(1 \times 10^{-1})33/[2.693 \times 10^3(9.2 \times 10^3)]$$
$$S = 0.0002$$
$$P_c = 1.51 \times 10^3(14)4.8 \times 10^{-2}/9.2 \times 10^3$$
$$P_c = 0.11 \text{ gpd/sq ft}$$

The time-drawdown graph for observation well 1 is presented in Figure 5.5.

Average values of transmissivity, storativity, and vertical hydraulic conductivity are 1500 gpd/ft, 0.0002, and 0.1 gpd/sq ft, respectively. The 20-hour test sampled a cylindrical volume of the aquifer with a height of 8 ft and a radius of 2000 ft. Well storage capacity impacts were appreciable until 36 minutes after pumping started and were taken into account in the analysis. Available geological data indicate that Dietrich Creek is not a recharge boundary; 14 ft of sandy clay separate the streambed and the aquifer. Steady-state conditions prevailed at the end of the test.

A 31-day pumping test was conducted in Case Study 5.5 using aquifer production and observation wells and aquitard and source bed observation wells (Neuman and Witherspoon, 1972, pp. 1292–1297). The production well fully penetrates the aquifer which underlies the city of Oxnard, California. An aquifer observation well was located 100 ft from the production well. Aquitard and source bed observation wells partially penetrate deposits. An upper aquitard observation well was located 62 ft from

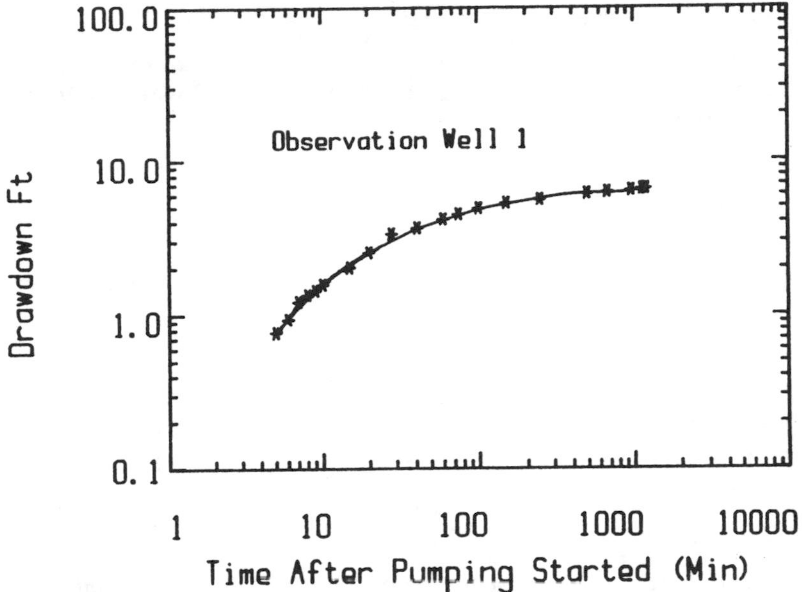

Figure 5.5. Time-drawdown graph for Case Study 5.4.

the production well and was open 22 ft above the top of the aquifer. A lower aquitard observation well was located 81 ft from the production well and was open 6 ft below the aquifer base. The upper source bed observation well was located 72 ft from the production well and the lower source bed observation well was located 100 ft from the production well.

The fine to coarse-grained sand and gravel aquifer with a thickness of 93 ft is encountered at a depth of 105 ft below land surface. The aquifer is overlain by a silty and sandy clay aquitard 45 ft in thickness, which in turn is overlain by a sand and gravel source bed. The aquifer is underlain by a silty and sandy clay aquitard 30 ft in thickness, which in turn is underlain by a source bed consisting of fine to coarse-grained sand and gravel. The storativities of the upper and lower aquitards were determined from laboratory consolidation tests and taken into account in pumping test analysis.

The constant discharge rate from the production well

Table 5.5. Database for Case Study 5.5

Time After Pumping Started (min)	Aquifer Adjusted Drawdown (ft)	Lower Aquitard Adjusted Drawdown (ft)
1	2.59	
3	3.61	
12	4.81	
25	5.36	
40	5.65	0.02
70	6.03	0.07
100	6.31	0.21
240	7.12	
510	7.73	2.33
860	8.02	
1210	8.29	4.82
2880	8.96	5.61
4700	9.43	
9200	9.81	6.14
Time After Pumping Started (min)	**Lower Source Bed Adjusted Drawdown (ft)**	**UpperAquitard Adjusted Drawdown (ft)**
510	0.06	
1210	0.72	
2880	1.98	
4700		0.02
9200	3.23	1.17

was 1000 gpm. Time-drawdown data for the aquifer observation well located 100 ft from the production well and data for aquitard observation wells for selected times are presented in Table 5.5.

The aquifer time-drawdown curve was matched to the appropriate model 4 type curve (Figure F.5 in Appendix F) and match point coordinates and the selected value of Gamma were used to calculate aquifer transmissivity and storativity and the vertical hydraulic conductivity of the lower aquitard. Particular attention was given to time-drawdown data before the effects of pumping reached the bottom of the lower aquitard and the observation wells in the upper aquitard. Calculated values of aquifer transmissivity and storativity and average lower aquitard vertical hydraulic conductivity and storativity are 130,000 gpd/ft, 1.11×10^{-4}, 2.88×10^{-2} gpd/sq ft, and 6.0×10^{-4}, respectively.

Drawdown data for the aquifer and aquitard observation wells were used to determine values of the ratio s_c/s at times after the effects of pumping reached aquitard observation wells. Values of vertical hydraulic conductivity calculated with ratios and Table E.5 in Appendix E and assigned to the bottom 21 ft of the upper aquitard, bottom 11 ft of the upper aquitard, and top 6 ft of the lower aquitard are 2.45×10^{-2}, 5.85×10^{-2}, and 4.17×10^{-2}, respectively. Calculations for the aquifer observation well and lower aquitard observation well are as follows:

Aquifer Observation Well

$$Q = 1000 \text{ gpm}, \quad r = 100 \text{ ft}, \quad m = 93 \text{ ft}, \quad m_c = 30$$

$$\text{Gamma} = 0.005$$

$$W(u,\text{Gamma}) = 0.1, \quad 1/u = 1000, \quad s = 0.088 \text{ ft}$$
$$t = 23 \text{ min}$$

$$T = 1.146 \times 10^2(1.0 \times 10^3)1.0 \times 10^{-1}/8.8 \times 10^{-2}$$
$$T = 1.30 \times 10^5 \text{ gpd/ft}$$

$$S = 1.3 \times 10^5(1.0 \times 10^{-3})23/[2.693 \times 10^3(1.0 \times 10^4)]$$
$$S = 1.11 \times 10^{-4}$$

$$P_c = 2.5 \times 10^{-5}(16)1.3 \times 10^5(1.11 \times 10^{-4})30]/$$
$$[1.0 \times 10^4(6.0 \times 10^{-4})]$$

$$P_c = 2.88 \times 10^{-2} \text{ gpd/sq ft}$$

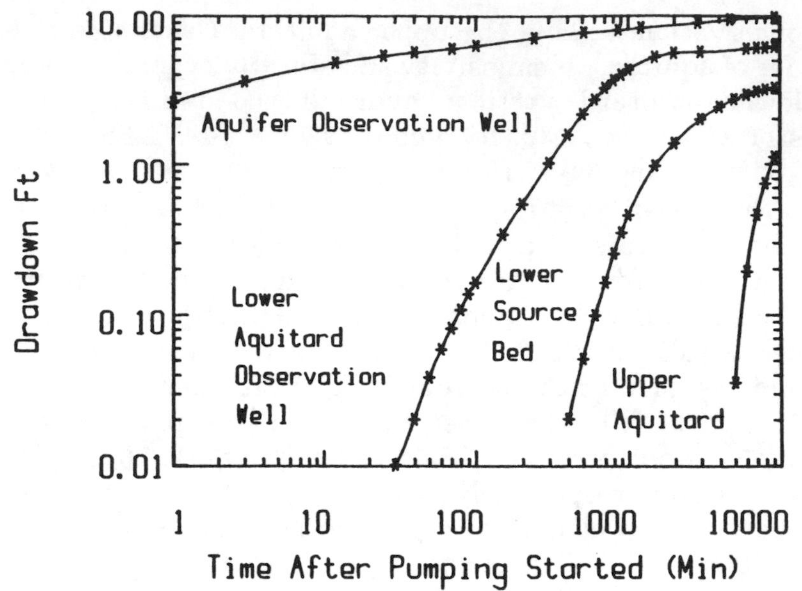

Figure 5.6. Time-drawdown graphs for Case Study 5.5.

Lower Aquitard Observation Well

$$r = 81 \text{ ft}, \quad z = 6 \text{ ft}, \quad t = 80 \text{ min}, \quad s_c = 7.8 \times 10^{2} \text{ ft}$$

$$s = 6.6 \text{ ft}, \quad s_c/s = 7.8 \times 10^{-2}/6.6, \quad s_c/s = 1.18 \times 10^{-2}$$

$$u = 2.693 \times 10^{3}(6.56 \times 10^{3})1.11 \times 10^{-4}/[1.3 \times 10^{5}(80)]$$

$$u = 1.89 \times 10^{-4}, \quad 1/u_c = 3.44 \times 10^{-1}$$

$$P_c = 2.693 \times 10^{3}(36)3.44 \times 10^{-1}(1.0 \times 10^{-4})/80$$

$$P_c = 4.17 \times 10^{-2} \text{ gpd/sq ft}$$

Time-drawdown graphs for observation wells are presented in Figure 5.6.

WATER TABLE AQUIFER SYSTEM

A pumping test was conducted in Case Study 5.6 on May 17 and 18, 1950, using a group of wells located about 4 miles southeast of the corporate limits of the city of Lawrenceville, Illinois (Prickett, 1965, pp. 5–14). Well logs indicate that the stratified aquifer consists of 100 ft of fine to medium sand with some coarse sand and much gravel. The 16-in.-diameter production well screen was in the lower 25 ft of the aquifer; observation wells 1, 3, 4, 6, 7, 8, 9, 10, and 11 have short screens in the upper 20 ft of the aquifer, and observation wells 2 and 5 were screened near the aquifer base. Distances to these observation wells from the production well were 12, 13, 100, 101, 200, 210, 215, 500, 502, 505, and 840 ft, respectively. Pumping was started at 8:38 a.m. on May 17 and was continued for a period of 24 hours at a constant rate of 1000 gpm until 9:10 a.m. on May 18. Time-drawdown data for observation well 5 and distance-drawdown data at the end of the test, adjusted for dewatering and partial penetration impacts, are presented in Table 5.6.

The time-drawdown curve for observation well 5 was found to be analogous to the Beta = 0.3 model 6 type curve (Figures F.6 and F.7 in Appendix F). Match point coordinates and the selected Beta value were used to calculate aquifer transmissivity, storativity, specific yield, and P_V/P_H ratio. The model 1 type curve (Figure F.2 in Appendix F) was matched to the distance-drawdown curve, and match point coordinates were used to calculate aquifer transmissivity and specific yield. Calculations are as follows:

Observation Well 5

$$Q = 1000 \text{ gpm}, \quad r = 200 \text{ ft}, \quad m = 100 \text{ ft}$$

$$\text{Match Pt. Coord. } W(u_A,\text{Beta}) = 0.1, \quad 1/u_A = 1.0$$
$$s = 4.1 \times 10^{-2} \text{ ft}$$

$$t = 9.0 \times 10^{-1} \text{ min} \quad \text{Beta} = 3.0 \times 10^{-1}$$

$$T = 1.146 \times 10^{2}(1.0 \times 10^{3})0.1/4.1 \times 10^{-2}$$
$$T = 279{,}512 \text{ gpd/ft}$$

Table 5.6. Database for Case Study 5.6

Time After Pumping Started (min)	Well 5 Adjusted Drawdown (ft)	Distance (ft)	Adjusted Drawdown (ft)
1	0.09	12	4.02
2	0.22	13	3.95
3	0.30	100	2.32
4	0.36	101	2.17
5	0.39	200	1.91
6	0.42	210	1.88
7	0.44	215	1.85
8	0.46	500	1.00
9	0.47	502	0.88
10	0.48	505	0.84
15	0.51	840	0.47
20	0.53		
30	0.56		
40	0.60		
50	0.66		
60	0.70		
70	0.75		
80	0.80		
90	0.83		
100	0.87		
150	1.01		
200	1.13		
300	1.26		
400	1.37		
500	1.42		
600	1.54		
700	1.60		
800	1.66		
900	1.70		
1000	1.74		
1480	1.91		

$$S = 2.79 \times 10^5(1)9.0 \times 10^{-1}/[2.693 \times 10^3(4.0 \times 10^4)]$$
$$S = 0.00233$$

$$P_V/P_H = 1.0 \times 10^4(3.0 \times 10^{-1})/4.0 \times 10^4$$
$$P_V/P_H = 7.5 \times 10^{-2} = 1/13$$

$$\text{Match Pt. Coord. } W(u_B,\text{Beta}) = 0.1, \qquad 1/u_B = 10$$
$$s = 4.3 \times 10^{-2} \text{ ft}$$
$$t = 1.28 \times 10^2 \text{ min}$$
$$T = 1.146 \times 10^2(1.0 \times 10^3)0.1/4.3 \times 10^{-2}$$
$$T = 266{,}511 \text{ gpd/ft}$$
$$S_y = 2.66 \times 10^5(0.1)1.28 \times 10^2/[2.693 \times 10^3(4.0 \times 10^4)]$$
$$S_y = 0.0316$$

Distance-Drawdown Graph

$$Q = 1000 \text{ gpm}, \qquad t = 1440 \text{ min}, \qquad m = 100 \text{ ft}$$
$$\text{Match Pt. Coord. } W(u) = 0.1, \qquad u = 0.1,$$
$$s = 4.4 \times 10^{-2}\ r^2 = 4.1 \times 10^5$$
$$T = 1.146 \times 10^2(1.0 \times 10^3)0.1/4.4 \times 10^{-2}$$
$$T = 260{,}454 \text{gpd/ft}$$
$$S_y = 2.60 \times 10^5(0.1)1.44 \times 10^3/[2.693 \times 10^3(4.1 \times 10^5)]$$
$$S_y = 0.0339$$

The time-drawdown graph for observation well 5 and the distance-drawdown graph are presented in Figures 5.7 and 5.8.

The average values of aquifer transmissivity, storativity, specific yield, and P_V/P_H ratio are 260,000 gpd/ft, 0.0023, 0.032, and 1/10, respectively. These values represent the aquifer within a radius of about 4650 ft from the production well. The specific yield is representative of the deposits above the cone of depression and may not be representative of the aquifer as a whole. Well storage capacity impacts were negligible before 1 minute of pumping, and partial penetration impacts were appreciable to a distance of 474 ft from the production well and were taken into account in the analysis. Delayed gravity yield impacts were appreciable until 691 minutes after pumping started.

A step drawdown test was conducted in Case Study 5.7 on April 27, 1954, using a production well near Granite City, Illinois (see Walton, 1962, pp. 63–64). The medium to fine grained sand with some gravel aquifer is 106 ft thick.

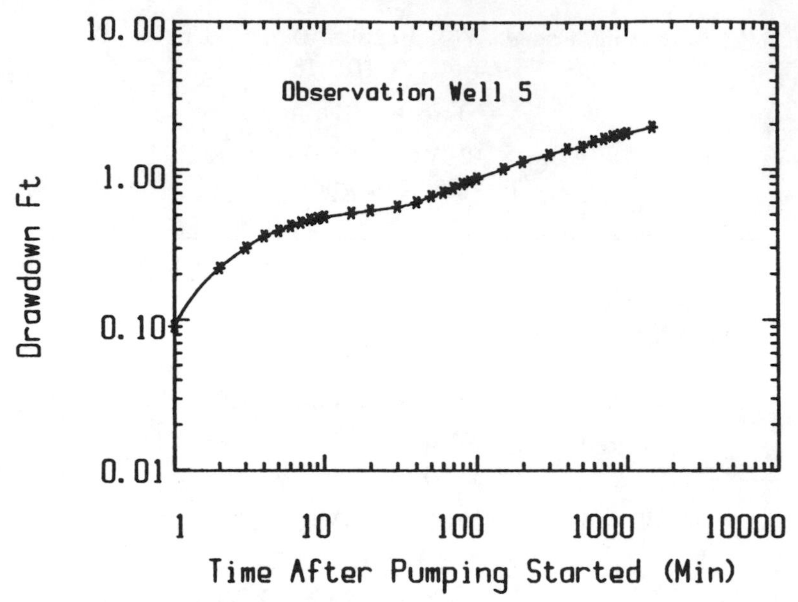

Figure 5.7. Time-drawdown graph for observation well 5 in Case Study 5.6.

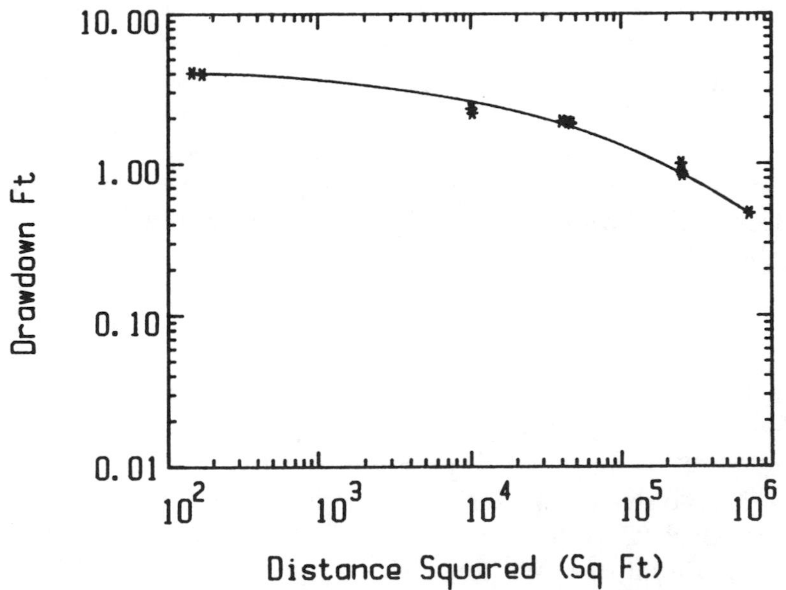

Figure 5.8. Distance-drawdown graph for Case Study 5.6.

Table 5.7. Database for Case Study 5.7

Time	Feet to Water	Pumping Rate (gpm)
9:23a.m.	23.95	0
9:50	29.13	1000
9:55	29.23	1000
10:00	29.30	1000
10:10	29.30	1000
10:15	29.32	1000
10:20	29.37	1000
10:30	29.35	1000
10:35	29.36	1000
10:52	29.40	1000
11:00	29.39	1000
11:25	29.43	1000
11:30	30.97	1260
11:50	31.01	1260
12:00	31.02	1260
12:15p.m.	31.04	1260
12:30	31.08	1260
12:45	31.08	1260
12:50	31.90	1400
1:00	31.95	1400
1:15	31.92	1400
1:30	31.90	1400
1:40	31.93	1400

The production well was screened opposite the lower 60 ft of the aquifer. The casing and screen had inside diameters of 30 in.

The test was started at 9:45 a.m. and was continued for 4 hours until 1:40 p.m. The discharge rates from the production well were 1000, 1280, and 1400 gpm. Test data are presented in Table 5.7.

The calculated values of well loss coefficient for steps 1 and 2 and step 2 and 3 are 0.04 sec^2/ft^5 and 0.11 sec^2/ft^5, respectively. Computations are as follows:

$$C_{1,2} = (1.59/0.62 - 5.43/2.22)/(2.22 + 0.62)$$

$$C_{1,2} = 0.04 \text{ sec}^2/\text{ft}^5$$

Table 5.8. Database for Case Study 5.8

Distance (ft)	Adjusted Drawdown (ft)
28	1.68
85	1.22
230	.83
420	.56

$$C_{2,3} = (0.72/0.27 - 1.59/0.62)/(0.62 + 0.27)$$

$$C_{2,3} = 0.11 \text{ sec}^2/\text{ft}^5$$

The average value of the well loss coefficient is 0.08 sec^2/ft^5. Well loss at the 1400 gpm rate was calculated to be 0.77 ft and amounted to about 10% of the total drawdown in the production well.

INDUCED STREAMBED INFILTRATION

A pumping test was conducted in Case Study 5.8 on February 20–23, 1956, using a group of six wells located along the Mad River about 4 miles northwest of the city of Springfield, Ohio (see Walton, 1963). Wells were located 400 ft west of Mad River. Observation wells were located 28, 85, 230, and 420 ft north from a production well. Logs of wells and test holes indicate that the 100-ft-thick aquifer consists of coarse sand and gravel, is about 4000 ft wide, and trends southwest to northeast through the pumping test site. The constant pumping rate was 1034 gpm and the test duration was 4310 minutes. Distance-drawdown data for observation wells at the end of the test when steady state conditions prevailed are presented in Table 5.8.

A straight line was drawn through the semilogarithmic distance-drawdown graph and aquifer transmissivity was calculated to be 547,000 gpd/ft. The recharging image well simulating induced streambed infiltration was located at a distance of 1372 ft from the production well. Aquifer

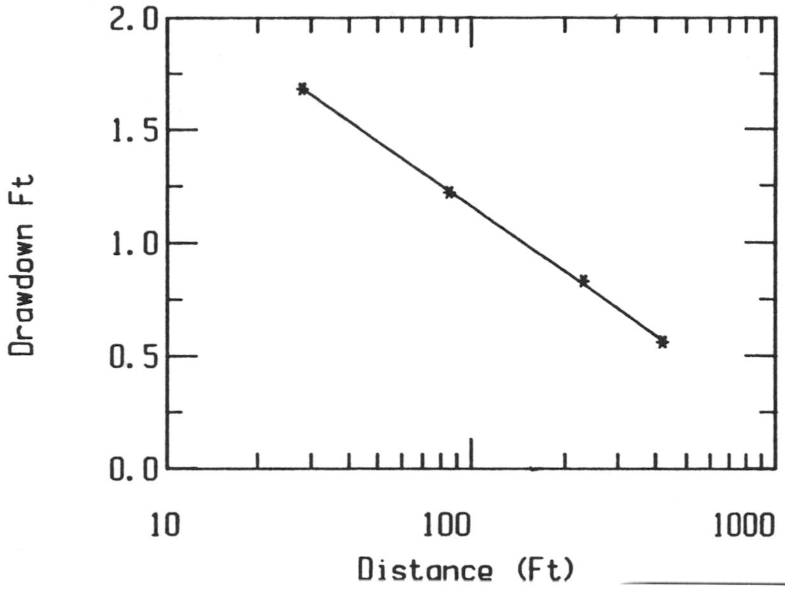

Figure 5.9. Distance-drawdown graph for Case Study 5.8.

specific yield is 0.01 and depletion of streamflow at the end of the test was 960 gpm or about 93% of the discharge rate. The reach of the streambed within the area of influence of pumping was calculated to be 8000 ft and the average width of the streambed is 82 ft. The streambed area of induced infiltration is 656,000 square ft. The induced infiltration rate is 9.2×10^4 gpd/acre. The average drawdown beneath the streambed was calculated to be 0.09 ft. The induced infiltration rate per ft of head loss is 1.0×10^6 gpd/acre/ft at a surface water temperature of 39°F. The streambed induced infiltration rate is in the clogged slow sand filter category (see Walton, 1970, p. 266). Well storage capacity and delayed gravity yield impacts were negligible at the end of the test. Calculations are as follows:

$$Q = 1034 \text{ gpm}, \qquad s_1 = 1.0 \text{ ft}$$
$$T = 528(1034)/1.0 \qquad T = 545{,}952 \text{ gpd/ft}$$

The distance-drawdown graph is presented in Figure 5.9.

APPENDIX A
Microcomputer Programs

Source codes for twelve IBM PC BASIC microcomputer programs (PT1–11 and CATALOG) are listed in this appendix. Program algorithms are documented as REM statements in listings to assist the reader in understanding the codes. Many statements in the programs could be restructured or eliminated to produce more efficient coding, but they have been intentionally written in their present form to provide clear explanations of processing steps.

PT1 PUMPING TEST DESIGN MODELS

```
10 CLEAR
20 CLS:KEY OFF
30 PRINT"Program: PT1"
40 PRINT"Version: IBM/PC 2.1"
50 PRINT"Model  : SIMULATION OF 1 OR 2-LAYER AQUIFER SYSTEM,"
60 PRINT"         UNIFORM PROPERTIES, WELL STORAGE CAPACITY,"
70 PRINT"         DELAYED GRAVITY YIELD, LEAKAGE, DEWATERING,"
80 PRINT"         RADIAL FLOW TO PRODUCTION WELL,FINITE-"
90 PRINT"         DIFFERENCE APPROXIMATION FOR PUMPING TEST"
100 PRINT"         DESIGN"
110 PRINT
120 PRINT"Program based in part on program presented"
130 PRINT"by Rushton,K.R. and S.C.Redshaw.1979.Seepage"
140 PRINT"and groundwater flow-numerical analysis by"
150 PRINT"analog and digital methods.John Wiley & Sons,Ltd."
160 PRINT"New York;  and Rathod,K.S and Rushton,K.R.1984."
170 PRINT"Numerical method of pumping test analysis using"
180 PRINT"microcomputers.GROUND WATER.Vol.22,No.5.":PRINT
190 PRINT"Press any key to continue"
200 T$=INKEY$:IF T$="" THEN 200
210 PRINT
220 PRINT"For details on equations and assumptions see"
230 PRINT"Walton,W.C.1985.2nd Edition.Practical aspects"
240 PRINT"of groundwater modeling.NATIONAL WATER WELL"
250 PRINT"ASSOC.Pages 388-391.":PRINT
260 DEFINT I,J,K,M,N:DEFDBL A-H,L,O-Z
270 PRINT"RESULTS WILL NOT APPEAR ON SCREEN"
280 PRINT"BUT WILL BE PRINTED ON PAPER":PRINT
290 PRINT"DATA BASE:":PRINT
300 LPRINT:LPRINT"DATA BASE:":LPRINT
310 INPUT"ENTER 1 FOR NONLEAKY,2 FOR LEAKY, OR 3 FOR WATER TABLE AQUIFER:";J2
320 INPUT"AQUIFER HORIZ. HYDR. CONDUCTIVITY (GPD/SQ FT)=";P
330 IF J2=1 THEN 410
340 IF J2=2 THEN 370
350 INPUT"AQUIFER VERT. HYDR. CONDUCTIVITY (GPD/SQ FT)=";PV
360 IF J2=3 THEN 380
370 INPUT"AQUITARD VERT. HYDR. CONDUCTIVITY (GPD/SQ FT)=";PC
380 INPUT"AQUIFER THICKNESS (FT)=";HF
390 IF J2=3 THEN 410
```

```
400 INPUT"AQUITARD THICKNESS (FT)=";HC
410 INPUT"ARTESIAN AQUIFER STORATIVITY (DIM)=";S1
420 'Transmissivity decreases due to dewatering are simulated
430 INPUT"WATER TABLE STORATIVITY (DIM)=";S2
440 INPUT"PRODUCTION WELL EFFECTIVE RADIUS (FT)=";R1
450 PRINT:PRINT"An infinite aquifer system is simulated"
460 PRINT
470 R9=100000!
480 LPRINT"AQUIFER HORIZ. HYDR. COND. (GPD/SQ FT)=" USING"######.##";P
490 IF J2=1 THEN 570
500 IF J2=2 THEN 530
510 LPRINT"AQUIFER VERT. HYDR. COND. (GPD/SQ FT)=" USING"#####.###";PV
520 IF J2=3 THEN 540
530 LPRINT"AQUITARD VERT. HYDR. COND. (GPD/SQ FT)=" USING"##.####^^^^";PC
540 LPRINT"AQUIFER THICKNESS (FT)=" USING"#####.##";HF
550 IF J2=3 THEN 570
560 LPRINT"AQUITARD THICKNESS (FT)=" USING"#####.##";HC
570 LPRINT"ARTESIAN AQUIFER STORATIVITY (DIM)=" USING"##.####^^^^";S1
580 LPRINT"WATER TABLE STORATIVITY (DIM)=" USING"##.####";S2
590 LPRINT"PRODUCT. WELL EFFECTIVE RADIUS (FT)=" USING"##.###";R1
600 PRINT:PRINT"Enter Y to revise data base"
610 INPUT"or N to continue";R$:PRINT
620 IF R$="Y" OR R$="y" THEN 290
630 IF R$<>"Y" AND R$<>"y" AND R$<>"N" AND R$<>"n" THEN 600
640 P=P/7.48:PC=PC/7.48:PV=PV/7.48
650 IF J2=1 THEN 740
660 IF J2=3 THEN 710
670 'Next statement calculates leakage coefficient
680 BL=P*HF/(PC/HC)
690 IF J2<>3 THEN 740
700 'Next statement calculates delayed gravity yield coefficient
710 AL=3*PV/(S2*HF)
720 'Establish logarithmic mesh:five mesh intervals per
730 'tenfold increase in radial distance
740 C=1.5848932#
750 R3=R1
760 N1=2
770 N1=N1+1
780 R3=R3*C
790 IF R3<R9 THEN 770
800 IF J2<>3 THEN 850
```

```
810 DIM Y(N1),X(N1)
820 FOR N=1 TO N1
830 Y(N)=0#
840 NEXT N
850 DIM R(50),R2(50),D(50),D1(50),S(50),H(50),Q(50)
860 DIM U(50),V(50),O1(5),O2(5)
870 'Calculate radial distances for mesh points
880 R(1)=R1/C
890 R2(1)=R(1)*R(1)
900 FOR N=2 TO N1
910 R(N)=C*R(N-1)
920 R2(N)=R(N)*R(N)
930 NEXT N
940 R(N1)=R9
950 R2(N1)=R9*R9
960 N2=N1-1
970 N3=N2-1
980 A=.460517#
990 A2=A*A
1000 INPUT"TOP OF AQUIFER DEPTH (FT)=";U1
1010 INPUT"BASE OF AQUIFER DEPTH (FT)=";L1
1020 INPUT"INITIAL WATER LEVEL DEPTH (FT)=";W
1030 Q1=0!
1040 'Initialize arrays
1050 FOR N=1 TO N1
1060 Q(N)=Q1
1070 D(N)=W
1080 D1(N)=W
1090 NEXT N
1100 LPRINT"TOP OF AQUIFER DEPTH (FT)=" USING"#####.##";U1
1110 LPRINT"BASE OF AQUIFER DEPTH (FT)=" USING"#####.##";L1
1120 LPRINT"INITIAL WATER LEVEL DEPTH (FT)=" USING"#####.##";W
1130 LPRINT"INFINITE AQUIFER SYSTEM":J1=2
1140 'The next statement sets observation well nodes
1150 'for display of time-drawdown data. Nodes may be
1160 'changed by reference to node numbering system
1170 'shown on distance-drawdown display at end of test
1180 O1(1)=13:O1(2)=15:O1(3)=17:O1(4)=19:O1(5)=21
1190 INPUT"PRODUCTION WELL DISCHARGE RATE (GPM)=";P1
1200 PRINT:PRINT"Drawdowns will be displayed at production"
1210 PRINT"well and five observation well sites at"
```

```
1220 PRINT"selected logarithmic spaced times including"
1230 PRINT"the end of the test. Then drawdowns"
1240 PRINT"will be displayed at end of test at"
1250 PRINT"selected logarithmic spaced distances":PRINT
1260 INPUT"TIME AFTER PUMPING STARTED AT END OF TEST (MIN)=";T9
1270 GOSUB 2390
1280 P1=P1*1440/7.48
1290 IF P1<0# THEN STOP
1300 P2=.125#*P1/(CDBL(ATN(1#))*A)
1310 GOSUB 2460
1320 T9=T9/1440
1330 I1=0
1340 T=0#
1350 'Set initial time step so that u<1.0 at well face
1360 T0=.25#*R2(2)*S1/(P*(L1-U1))
1370 IF T0>.000001# THEN T0=.000001#
1380 'Calculations for each time step
1390 T=T+T0
1400 IF T<T9 THEN 1440
1410 T0=T9-(T-T0)
1420 T=T9
1430 I1=100
1440 IF J2<>3 THEN 1580
1450 'Simulate delayed gravity yield
1460 F=AL*T0
1470 IF F>100! THEN 1490
1480 FA=EXP(-F)
1490 FB=1!-FA
1500 FC=FB/(AL*T0)
1510 FOR N=1 TO N1
1520 X(N)=FA*Y(N)
1530 Q(N)=AL*S2*X(N)
1540 NEXT N
1550 IF J2<>3 THEN 1580
1560 K2=4
1570 GOTO 1600
1580 K2=1
1590 'Select appropriate storativity
1600 FOR K=1 TO K2
1610 FOR N=1 TO N2
1620 Z=L1-.5#*(D(N)+D(N+1))
```

```
1630 IF J2=3 THEN 1670
1640 S3=S2
1650 GOTO 1680
1660 'Simulate delayed gravity yield
1670 S3=S1+FB*S2
1680 IF Z<(L1-U1) OR J2=3 THEN 1710
1690 Z=L1-U1
1700 S3=S1
1710 H(N)=A2/(Z*P)
1720 IF J2<>2 THEN 1750
1730 'Next statement simulates leaky artesian conditions
1740 Q(N)=P*Z*(D1(N)-W)/BL
1750 S(N)=T0/(S3*R2(N))
1760 NEXT N
1770 'Modify coefficients to take into account
1780 'well storage and condition near well face
1790 H(1)=.0001#*H(1)
1800 S(1)=2#*T0*A/R2(2)
1810 S(2)=2#*S(2)
1820 'Modify coefficients for outer boundary condition
1830 H(N2)=(LOG(R(N1)/R(N2)))*(LOG(R(N1)/R(N2)))/(Z*P)
1840 H(N1)=10000000000#
1850 S(N2)=2#*T0*A/((R(N1)-R(N2-1))*S3*R(N2))
1860 S(N1)=2#*T0*A/((R(N1)-R(N2))*S3*R(N1))
1870 IF J1=1 THEN S(N1)=.000000001#*S(N1)
1880 'Gaussian elimination
1890 U(1)=1#/H(1)+1#/S(1)
1900 V(1)=D1(1)/S(1)+P2
1910 FOR N=2 TO N2
1920 CLS:PRINT"COMPUTATIONS ARE IN PROGRESS"
1930 U(N)=1#/H(N-1)+1#/H(N)+1#/S(N)
1940 U(N)=U(N)-(1#/H(N-1))*(1#/H(N-1))/U(N-1)
1950 RQ2=R2(N)*Q(N)
1960 V(N)=D1(N)/S(N)-RQ2+(1#/H(N-1)*V(N-1))/U(N-1)
1970 NEXT N
1980 RQ3=.5#*R(N2)*(R(N1)-R(N3))*Q(N2)/A
1990 V(N2)=D1(N2)/S(N2)-RQ3
2000 V(N2)=V(N2)+(V(N3)/H(N3))/U(N3)
2010 U(N1)=1#/H(N2)+1#/S(N1)
2020 U(N1)=U(N1)-(1#/H(N2))*(1#/H(N2))/U(N2)
2030 RQ4=.5#*R(N1)*(R(N1)-R(N2))*Q(N1)/A
```

```
2040 V(N1)=D1(N1)/S(N1)-RQ4
2050 V(N1)=V(N1)+(V(N2)/H(N2))/U(N2)
2060 D(N1)=V(N1)/U(N1)
2070 FOR J=1 TO N2
2080 N=N2-J+1
2090 D(N)=(V(N)+1#/H(N)*D(N+1))/U(N)
2100 NEXT J
2110 'If drawdown in production well below top of aquifer
2120 'reaches more than 90% of aquifer thickness then
2130 'well runs dry and program stops
2140 IF D(1)<(.9#*L1+.1#*U1) THEN 2180
2150 LPRINT"EXCESSIVE DRAWDOWN"
2160 GOSUB 2540
2170 STOP
2180 NEXT K
2190 'Drawdown at 5 observation sites
2200 FOR M=1 TO 5
2210 K1=O1(M)
2220 O2(M)=D(K1)
2230 NEXT M
2240 IF T*1440>.1 THEN LPRINT USING"######.##  ";T*1440,D(2),O2(1),O2(2),O2(3),O
2(4),O2(5)
2250 FOR N=1 TO N1
2260 D1(N)=D(N)
2270 NEXT N
2280 IF J2<>3 THEN 2350
2290 'Transfer values of drawdowns to old drawdowns
2300 FOR N=1 TO N1
2310 Y(N)=X(N)+FC*(D(N)-D1(N))
2320 NEXT N
2330 'End of calculations for one time step
2340 'Calculate new time step
2350 T0=T*.58489324
2360 IF I1=0 THEN 1390
2370 GOSUB 2540
2380 END
2390 LPRINT:LPRINT"COMPUTATION RESULTS:":LPRINT
2400 LPRINT"PRODUCTION WELL DISCHARGE RATE (GPM)=" USING"#####.##";P1
2410 LPRINT
2420 LPRINT"TIME-DRAWDOWN OR WATER LEVEL VALUES (FT)"
2430 LPRINT
2440 LPRINT"                              SELECTED DISTANCES (FT)":LPRINT
```

```
2450 RETURN
2460 FOR M=1 TO 5
2470 K1=O1(M)
2480 O2(M)=R(K1)
2490 NEXT M
2500 LPRINT"TIME(MIN)  ";
2510 LPRINT USING"######.##  ";R(2),O2(1),O2(2),O2(3),O2(4),O2(5)
2520 LPRINT
2530 RETURN
2540 LPRINT
2550 LPRINT"TIME AFTER PUMPING STARTED(MIN)=" USING"#####.##";T9*1440
2560 LPRINT
2570 LPRINT"DISTANCE-DRAWDOWN OR WATER LEVEL VALUES AT END OF PUMPING PERIOD"
2580 LPRINT
2590 LPRINT"NODE  RADIUS(FT)   DRAWDOWN OR WATER LEVEL (FT)"
2600 LPRINT"  NO"
2610 FOR N=2 TO N1
2620 IF D1(N)<.001 OR D1(N)<(W+.001) THEN 2650
2630 LPRINT USING"###";N,
2640 LPRINT USING"   #####.##    ";R(N),D1(N)
2650 NEXT N
2660 RETURN
```

PT2 PARTIAL PENETRATION WELL FUNCTION

```
10 CLEAR
20 CLS:KEY OFF
30 AF$="##.####^^^^"
40 PRINT"Program: PT2"
50 PRINT"Author : W.C. Walton"
60 PRINT"Version: IBM/PC 2.1; Copyright 1987 Lewis Publishers, Inc."
70 PRINT"Purpose: COMPUTE W(U,R/M(PV/PH)^.5/M,LP/M,DP/M,LO/M,DO/M)"
80 PRINT"         AND PARTIAL PENETRATION IMPACTS FOR PUMPING TEST"
90 PRINT"         DESIGN"
100 PRINT
110 LPRINT
120 PRINT"Program based in part on equations given by"
130 PRINT"Hantush,M.S.1964.Hydraulics of wells.In Advances"
140 PRINT"In Hydroscience.Editor,Ven Te Chow.Vol. 1."
150 PRINT"Academic Press,Inc.New York":PRINT
160 PRINT"For details on equations and assumptions see"
170 PRINT"Walton,W.C.1985.2nd edition.Practical aspects of"
180 PRINT"groundwater modeling.NATIONAL WATER WELL ASSOC."
190 PRINT"Pages 163-170,339":PRINT
200 DDP=0!:DPP=0!
210 PRINT"DATA BASE:":PRINT
220 LPRINT"DATA BASE:":LPRINT
230 PRINT"Enter Y to calculate partial penetration"
240 INPUT"impacts or N to continue";PP$:PRINT
250 IF PP$<>"Y" AND PP$<>"y" AND PP$<>"N" AND PP$<>"n" THEN 230
260 IF PP$="N" OR PP$="n" THEN 350
270 INPUT"PRODUCTION WELL DISCHARGE RATE (GPM)=";Q
280 LPRINT"PRODUCTION WELL DISCHARGE RATE (GPM)=" USING"#####.##";Q
290 INPUT"AQUIFER STORATIVITY (DIM)=";STOR
300 LPRINT"AQUIFER STORATIVITY (DIM)=" USING AF$;STOR
310 INPUT"TIME AFTER PUMPING STARTED (MIN)=";TIME
320 LPRINT"TIME AFTER PUMPING STARTED (MIN)=" USING"#####.##";TIME
330 INPUT"DRAWDOWN WITH FULL PENETRATION (FT)=";DFP
340 LPRINT"DRAWDOWN WITH FULL PENETRATION (FT)=" USING"####.##";DFP
350 Z=0
360 INPUT"AQUIFER HORIZ. HYDR. CONDUCTIVITY (GPD/SQ FT)=";PH
370 LPRINT"AQUIFER HORIZ. HYDR. CONDUCTIVITY GPD/SQ FT)=" USING AF$;PH
380 INPUT"AQUIFER VERT. HYDR. CONDUCTIVITY (GPD/SQ FT)=";PV
```

```
390 LPRINT"AQUIFER VERT. HYDR. CONDUCTIVITY (GPD/SQ FT)=" USING AF$;PV
400 INPUT"RADIAL DISTANCE TO WELL (FT)=";RD
410 LPRINT"RADIAL DISTANCE TO WELL (FT)=" USING AF$;RD
420 INPUT"AQUIFER THICKNESS (FT)=";MA
430 LPRINT"AQUIFER THICKNESS (FT)=" USING AF$;MA
440 INPUT"DISTANCE FROM AQUIFER TOP TO BOTTOM OF PROD. WELL (FT)=";LP
450 LPRINT"DIST. FROM AQUIFER TOP TO BOTTOM OF PROD. WELL (FT)=" USING AF$;LP
460 INPUT"DIST. FROM AQUIFER TOP TO TOP OF PROD.WELL SCREEN (FT)=";DP
470 LPRINT"DIST. FROM AQUIFER TOP TO PROD. WELL SCREEN TOP (FT)" USING AF$;DP
480 INPUT"DISTANCE FROM AQUIFER TOP TO BOTTOM OF OBS. WELL (FT)=";LL
490 LPRINT"DIST. FROM AQUIFER TOP TO BOTTOM OF OBS. WELL (FT)=" USING AF$;LL
500 INPUT"DIST. FROM AQUIFER TOP TO TOP OF OBS. WELL SCREEN (FT)=";DD
510 LPRINT"DIST. FROM AQUIFER TOP TO TOP OF OBS. WELL SCREEN(FT)=" USING AF$;DD
520 IF PP$="Y" OR PP$="y" THEN 560
530 INPUT"U=";U
540 LPRINT"U=" USING AF$;U
550 GOTO 570
560 U=2693*RD^2*STOR/(PH*MA*TIME)
570 LPRINT
580 'Calculate W(partial penetration)
590 FOR K=1 TO 50
600 CLS:PRINT"COMPUTATIONS ARE IN PROGRESS"
610 S=K*3.1416*RD*(PV/PH)^.5/MA
620 GOSUB 920
630 GOSUB 870
640 NEXT K
650 WUPAR=2*MA^2*Z/(3.1416^2*(LP-DP)*(LL-DD))
660 CLS
670 PRINT"COMPUTATION RESULTS:"
680 LPRINT"COMPUTATION RESULTS:"
690 PRINT:PRINT"WELL FUNCTION=" USING AF$;WUPAR
700 LPRINT:LPRINT"WELL FUNCTION=" USING AF$;WUPAR
710 IF PP$="N" OR PP$="n" THEN 780
720 DPP=DFP+114.6*Q*WUPAR/(PH*MA)
730 DDP=114.6*Q*WUPAR/(PH*MA)
740 PRINT"DRAWDOWN DUE TO PART. PENETR. IMPACTS (FT)=" USING "####.##";DDP
750 PRINT"DRAWDOWN WITH PART. PENETR. (FT)=" USING"####.##";DPP
760 LPRINT"DRAWDOWN DUE TO PART. PENETR. IMPACTS (FT)=" USING"####.##";DDP
770 LPRINT"DRAWDOWN WITH PART. PENETR. (FT)=" USING"####.##";DPP
780 LPRINT:PRINT
790 PRINT"Enter Y for another computation with a"
```

```
800 INPUT"new data base or N to end program";A$
810 IF A$<>"Y" AND A$<>"y" AND A$<>"N" AND A$<>"n" THEN 790
820 IF A$="N" OR A$="n" THEN 850
830 LPRINT:PRINT
840 GOTO 10
850 END
860 'Subroutine to compute sum of sin factors
870 AA=SIN(K*3.1416*(LP/MA))-SIN(K*3.1416*(DP/MA))
880 BB=SIN(K*3.1416*(LL/MA))-SIN(K*3.1416*(DD/MA))
890 Z=Z+1/K^2*(AA*BB*WS)
900 RETURN
910 'Subroutine to compute W(u,B)
920 IF U>5! THEN WS=0:GOTO 1400
930 IF S>2 THEN 1300
940 IF U>=1 THEN 970
950 WU=-LOG(U)-.5772156600000009#+.9999919300000017#*U-.24991055#*U*U
960 WU=WU+.0551997*U^3-9.76004E-03*U^4+1.07857E-03*U^5:GOTO 1030
970 WU=U^4+8.573328740100025#*U^3+18.059016973#*U^2+8.634760892499999#*U
980 WU=WU+.2677737343#
990 WUP=U^4+9.573322345400009#*U^3+25.6329561486#*U^2+21.0996530827#*U
1000 WUP=WUP+3.958496922800001#
1010 WU=WU/WUP
1020 WU=WU/(U*EXP(U))
1030 L=S/3.75
1040 V=1+3.5156229#*L^2+3.0899424#*L^4+1.2067492#*L^6+.265973*L^8+.0360768*L^10
1050 V=V+.0045813*L^12
1060 F=S/2
1070 H=-LOG(F)*V-.5772156600000009#+.422784*F^2+.23069756#*F^4+.0348859*F^6
1080 H=H+2.62698E-03*F^8+.0001075*F^10+.0000074*F^12
1090 IF S=0 THEN WS=WU:GOTO 1400
1100 N=S^2/(4*U)
1110 IF N>5 THEN WS=2*H:GOTO 1400
1120 IF U<=.9 THEN 1150
1130 A=U+.5858:B=U+3.414:C=S*S/4
1140 WS=1.5637*EXP(-A-C/A)/A+4.54*EXP(-B-C/B)/B:GOTO 1400
1150 IF U<.05 THEN 1190
1160 IF U>S/2 THEN 1180
1170 C=-(1.75*U)^-.488*S:WS=2*H-4.8*10^C:GOTO 1400
1180 WS=WU-(S/(4.7*U^.6))^2:GOTO 1400
1190 IF U>.01 AND S<.1 THEN 1180
1200 IF N<1 THEN 1270
```

```
1210 WN=N^4+8.5733287401000025#*N^3+18.059016973#*N^2+8.634760892499999#*N
1220 WN=WN+.267773734300001#
1230 WNP=N^4+9.573322345400009#*N^3+25.6329561486#*N^2+21.0996530827#*N
1240 WNP=WNP+3.958496922800012#
1250 WN=WN/WNP
1260 WN=WN/(N*EXP(N)):GOTO 1290
1270 WN=-LOG(N)+.999991930000017#*N-.577215660000009#-.24991055#*N^2
1280 WN=WN+.0551997*N^3-9.76004E-03*N^4+1.07857E-03*N^5
1290 WS=2*H-WN*V:GOTO 1400
1300 M=-((S-2*U)/(2*U^.5))
1310 IF M<0 THEN M=ABS(M):GOSUB 1330:GOTO 1390
1320 GOSUB 1330:GOTO 1380
1330 NP=1+.0705230784#*M+.0422820123#*M^2+9.270527200000026D-03*M^3
1340 NP=NP+1.52014E-04*M^4+2.76567E-04*M^5+4.30638E-05*M^6
1350 IF NP>100! THEN 1370
1360 N=1/NP^16:RETURN
1370 N=0!:RETURN
1380 WS=(3.1416/(2*S))^.5*EXP(-S)*N:GOTO 1400
1390 WS=(3.1416/(2*S))^.5*EXP(-S)*(2-N)
1400 RETURN
```

PT3 WELL FUNCTION VALUE INTERPOLATION

```
10 CLS:CLEAR:KEY OFF
20 AF$="###.####":AD$="##.####^^^^"
30 PRINT"Program: PT3"
40 PRINT"Author : W.C. Walton"
50 PRINT"Version: IBM/PC 2.1"
60 PRINT"Purpose: INTERPRET WELL FUNCTION VALUES"
70 PRINT"         BETWEEN VARIABLE VALUES GIVEN"
80 PRINT"         IN A TABLE FOR PUMPING TEST"
90 PRINT"         ANALYSIS"
100 PRINT:PRINT"Program based in part on program"
110 PRINT"presented by Poole,L.,M.Borchers,and"
120 PRINT"K.Koessel.1981.Some Common BASIC programs."
130 PRINT"OSBORNE/McGraw-Hill.pp. 84-85":PRINT
140 DEFDBL A-Z:DEFINT I,J
150 DIM X(6),Y(6)
160 N=6
170 PRINT:PRINT"DATA BASE:":PRINT
180 LPRINT"DATA BASE:":LPRINT
190 PRINT"Enter data for 3 known points preceeding"
200 PRINT"interpolation point and data for 3 known"
210 PRINT"points following interpolation point":PRINT
220 FOR I=1 TO N
230 PRINT"POINT NUMBER=";I
240 LPRINT"POINT NUMBER=";I
250 INPUT"VARIABLE VALUE OF POINT=";X(I)
260 LPRINT"VARIABLE VALUE OF POINT=" USING AD$;X(I)
270 INPUT"WELL FUNCTION VALUE OF POINT=";Y(I)
280 LPRINT"WELL FUNCTION VALUE OF POINT=" USING AF$;Y(I)
290 NEXT I
300 PRINT:INPUT"VARIABLE VALUE OF INTERPOLATION POINT=";XX
310 LPRINT:LPRINT"VARIABLE VALUE OF INTERPOLATION POINT=" USING AD$;XX
320 YY=0!
330 'Solve Lagrange interpolation equation
340 FOR J=1 TO N
350 TT=1!
360 FOR I=1 TO N
370 IF I=J THEN 390
380 TT=TT*(XX-X(I))/(X(J)-X(I))
```

```
390 NEXT I
400 YY=YY+TT*Y(J)
410 NEXT J
420 PRINT:PRINT"COMPUTATION RESULTS:"
430 LPRINT:LPRINT"COMPUTATION RESULTS:"
440 PRINT:PRINT"WELL FUNCTION VALUE OF INTERPOLATION POINT=" USING AF$;YY
450 LPRINT:LPRINT"WELL FUNCTION VALUE OF INTERPOLATION POINT=" USING AF$;YY
460 PRINT:PRINT"Enter Y for another calculation or"
470 INPUT"N to end program";A$
480 IF A$<>"Y" AND A$<>"y" AND A$<>"N" AND A$<>"n" THEN 460
490 IF A$="Y" OR A$="y" THEN 10
500 END
```

PT4 W(u) WELL FUNCTION

```
10 CLS:CLEAR:KEY OFF
20 AF$="##.####^^^^"
30 PRINT"Program: PT4"
40 PRINT"Author : W.C. Walton"
50 PRINT"Version: IBM/PC 2.1; Copyright 1987 Lewis Publishers, Inc."
60 PRINT"Purpose: CALCULATE W(U) + W(U,R/M(PV/PH)^.5,L/M,D/M,LO/M,DO/M)"
70 PRINT"         FOR PUMPING TEST ANALYSIS"
80 PRINT:PRINT"Program based in part on polynomial approximation"
90 PRINT"given by Abramowitz,M. and I.A.Stegun.1970.Handbook"
100 PRINT"of mathematical functions.Dover Publications.New York"
110 PRINT:PRINT"For details on equations and assumptions see"
120 PRINT"Walton,W.C.1985.2nd edition.Practical aspects of"
130 PRINT"groundwater modeling.NATIONAL WATER WELL ASSOC."
140 PRINT"Pages 163-170,337-339":PRINT
150 'enter data base
160 PRINT"DATA BASE:":LPRINT:LPRINT"DATA BASE:"
170 PRINT:INPUT"U=";U
180 LPRINT:LPRINT"U=" USING AF$;U:LPRINT
190 'Calculate W(u) using polynomial approximation
200 A=U^2:B=U^3:C=U^4:D=U^5
210 IF U>1 THEN 240
220 W=-LOG(U)-.5772156600000006#+.9999919300000007#*U-.24991055#*A+.0551997*B-9.
76004E-03*C+1.07857E-03*D
230 GOTO 270
240 W=C+8.573328740100012#*B+18.059016973#*A+8.634760892499999#*U+.2677737343000
006#
250 W=W/(C+9.573322345400008#*B+25.6329561486#*A+21.0996530827#*U+3.958496922800
006#)
260 W=W/(U*EXP(U))
270 PRINT:PRINT"Enter Y for partially penetrating wells"
280 INPUT"or N for fully penetrating wells";AP$
290 IF AP$<>"Y" AND AP$<>"y" AND AP$<>"N" AND AP$<>"n" THEN 270
300 IF AP$="N" OR AP$="n" THEN 650
310 PRINT:PRINT"Program based in part on equations given by"
320 PRINT"Hantush,M.S.1964.Hydraulics of wells.In Advances"
330 PRINT"In Hydroscience.Editor,Ven Te Chow.Vol. 1."
340 PRINT"Academic Press,Inc.New York":PRINT
350 'Enter partial penetration sub-data base
360 Z=0
```

```
370 INPUT"AQUIFER HORIZ. HYDR. CONDUCTIVITY (GPD/SQ FT)=";PH
380 LPRINT"AQUIFER HORIZ. HYDR. CONDUCTIVITY (GPD/SQ FT)=" USING AF$;PH
390 INPUT"AQUIFER VERT. HYDR. CONDUCTIVITY (GPD/SQ FT)=";PV
400 LPRINT"AQUIFER VERT. HYDR. CONDUCTIVITY (GPD/SQ FT)=" USING AF$;PV
410 INPUT"RADIAL DISTANCE TO WELL (FT)=";RD
420 LPRINT"RADIAL DISTANCE TO WELL (FT)=" USING AF$;RD
430 INPUT"AQUIFER THICKNESS (FT)=";MA
440 LPRINT"AQUIFER THICKNESS (FT)=" USING AF$;MA
450 INPUT"DISTANCE FROM AQUIFER TOP TO BOTTOM OF PROD. WELL (FT)=";LP
460 LPRINT"DIST. FROM AQUIFER TOP TO BOTTOM OF PROD. WELL (FT)=" USING AF$;LP
470 INPUT"DIST. FROM AQUIFER TOP TO TOP OF PROD.WELL SCREEN (FT)=";DP
480 LPRINT"DIST. FROM AQUIFER TOP TO PROD. WELL SCREEN TOP (FT)" USING AF$;DP
490 INPUT"DISTANCE FROM AQUIFER TOP TO BOTTOM OF OBS. WELL (FT)=";LL
500 LPRINT"DIST. FROM AQUIFER TOP TO BOTTOM OF OBS. WELL (FT)=" USING AF$;LL
510 INPUT"DIST. FROM AQUIFER TOP TO TOP OF OBS. WELL SCREEN (FT)=";DD
520 LPRINT"DIST. FROM AQUIFER TOP TO TOP OF OBS. WELL SCREEN(FT)=" USING AF$;DD
530 LPRINT
540 'Calculate W(u,r/m(PV/PH)^.5,l/m,d/m,lo/m,do/m) using polynomial
550 'approximation and summation
560 FOR K=1 TO 50
570 CLS:PRINT"COMPUTATIONS ARE IN PROGRESS"
580 S=K*3.1416*RD*(PV/PH)^.5/MA
590 GOSUB 850
600 GOSUB 800
610 NEXT K
620 WUPAR=2*MA^2*Z/(3.1416^2*(LP-DP)*(LL-DD))
630 WUPT=W+WUPAR
640 GOTO 660
650 WUPT=W
660 CLS
670 PRINT:PRINT"COMPUTATION RESULTS:"
680 LPRINT:LPRINT"COMPUTATION RESULTS:"
690 PRINT:PRINT"WELL FUNCTION=" USING AF$;WUPT
700 LPRINT:LPRINT"WELL FUNCTION=" USING AF$;WUPT
710 LPRINT:PRINT
720 PRINT"Enter Y for another computation"
730 INPUT"or N to end program";A$
740 IF A$<>"Y" AND A$<>"y" AND A$<>"N" AND A$<>"n" THEN 720
750 IF A$="N" OR A$="n" THEN 780
760 LPRINT:PRINT
770 GOTO 10
```

```
780 END
790 'Subroutine to compute sum of sin factors
800 AA=SIN(K*3.1416*(LP/MA))-SIN(K*3.1416*(DP/MA))
810 BB=SIN(K*3.1416*(LL/MA))-SIN(K*3.1416*(DD/MA))
820 Z=Z+1/K^2*(AA*BB*WS)
830 RETURN
840 'Subroutine to compute W(u,B)
850 IF U>5! THEN WS=0:GOTO 1330
860 IF S>2 THEN 1230
870 IF U>=1 THEN 900
880 WU=-LOG(U)-.57721566000000009#+.99999193000000017#*U-.24991055#*U*U
890 WU=WU+.0551997*U^3-9.76004E-03*U^4+1.07857E-03*U^5:GOTO 960
900 WU=U^4+8.573328740100025#*U^3+18.059016973#*U^2+8.634760892499999#*U
910 WU=WU+.2677737343#
920 WUP=U^4+9.573322345400009#*U^3+25.6329561486#*U^2+21.0996530827#*U
930 WUP=WUP+3.95849692280001#
940 WU=WU/WUP
950 WU=WU/(U*EXP(U))
960 L=S/3.75
970 V=1+3.5156229#*L^2+3.0899424#*L^4+1.2067492#*L^6+.265973*L^8+.0360768*L^10
980 V=V+.0045813*L^12
990 F=S/2
1000 H=-LOG(F)*V-.57721566000000009#+.422784*F^2+.23069756#*F^4+.0348859*F^6
1010 H=H+2.62698E-03*F^8+.0001075*F^10+.0000074*F^12
1020 IF S=0 THEN WS=WU:GOTO 1330
1030 N=S^2/(4*U)
1040 IF N>5 THEN WS=2*H:GOTO 1330
1050 IF U<=.9 THEN 1080
1060 A=U+.5858:B=U+3.414:C=S*S/4
1070 WS=1.5637*EXP(-A-C/A)/A+4.54*EXP(-B-C/B)/B:GOTO 1330
1080 IF U<.05 THEN 1120
1090 IF U>S/2 THEN 1110
1100 C=-(1.75*U)^-.488*S:WS=2*H-4.8*10^C:GOTO 1330
1110 WS=WU-(S/(4.7*U^.6))^2:GOTO 1330
1120 IF U>.01 AND S<.1 THEN 1110
1130 IF N<1 THEN 1200
1140 WN=N^4+8.573328740100025#*N^3+18.059016973#*N^2+8.634760892499999#*N
1150 WN=WN+.267773734300001#
1160 WNP=N^4+9.573322345400009#*N^3+25.6329561486#*N^2+21.0996530827#*N
1170 WNP=WNP+3.958496922800012#
1180 WN=WN/WNP
```

```
1190 WN=WN/(N*EXP(N)):GOTO 1220
1200 WN=-LOG(N)+.99999193000000017#*N-.57721566000000009#-.24991055#*N^2
1210 WN=WN+.0551997*N^3-9.76004E-03*N^4+1.07857E-03*N^5
1220 WS=2*H-WN*V:GOTO 1330
1230 M=-((S-2*U)/(2*U^.5))
1240 IF M(0 THEN M=ABS(M):GOSUB 1260:GOTO 1320
1250 GOSUB 1260:GOTO 1310
1260 NP=1+.0705230784#*M+.0422820123#*M^2+9.2705272000000026D-03*M^3
1270 NP=NP+1.52014E-04*M^4+2.76567E-04*M^5+4.30638E-05*M^6
1280 IF NP)100! THEN 1300
1290 N=1/NP^16:RETURN
1300 N=0!:RETURN
1310 WS=(3.1416/(2*S))^.5*EXP(-S)*N:GOTO 1330
1320 WS=(3.1416/(2*S))^.5*EXP(-S)*(2-N)
1330 RETURN
```

PT5 W(u,r/B) WELL FUNCTION

```
10 CLS:CLEAR:KEY OFF
20 AF$="##.####^^^^"
30 PRINT"Program: PT5"
40 PRINT"Author : W.C. Walton"
50 PRINT"Version: IBM/PC 2.1; Copyright 1987 Lewis Publishers, Inc."
60 PRINT"Purpose: CALCULATE W(u,r/B)"
70 PRINT"         + W(u,r/m(PV/PH)^.5,l/m,d/m,lo/m,do/m)
80 PRINT"         FOR PUMPING TEST ANALYSIS"
90 PRINT:PRINT"Program based in part on code presented by"
100 PRINT"Sandberg,R.,R.B.Scheibach,D.Koch and T.A.Prickett."
110 PRINT"1981.Selected hand-held calculator codes for the"
120 PRINT"evaluation of the probable cumulative hydrologic"
130 PRINT"impacts of mining.U.S.Dept. of Interior,Office of"
140 PRINT"Surface Mining.H-D3004/030-81-1029F"
150 PRINT:PRINT"For details on equations and assumptions see"
160 PRINT"Walton,W.C.1985.2nd edition.Practical aspects of"
170 PRINT"groundwater modeling.NATIONAL WATER WELL ASSOC."
180 PRINT"Pages 163-170,337-339":PRINT
190 'enter data base
200 PRINT:PRINT"DATA BASE:":LPRINT:LPRINT"DATA BASE:":LPRINT
210 PRINT:INPUT"U=";U
220 INPUT"R/B=";S
230 LPRINT"U="USING AF$;U
240 LPRINT"R/B="USING AF$;S
250 IF U>5! THEN WS=0:GOTO 730
260 IF S>2 THEN 650
270 'Calculate W(u) using polynomial approximation
280 IF U>=1 THEN 310
290 WU=-LOG(U)-.57721566000000009#+.99999193000000017#*U-.24991055#*U*U
300 WU=WU+.0551997*U^3-9.76004E-03*U^4+1.07857E-03*U^5:GOTO 380
310 WU=U^4+8.5733287401000025#*U^3+18.059016973#*U^2+8.6347608924999999#*U
320 WU=WU+.2677737343#
330 WUP=U^4+9.5733223454000009#*U^3+25.6329561486#*U^2+21.0996530827#*U
340 WUP=WUP+3.9584969228000001#
350 WU=WU/WUP
360 WU=WU/(U*EXP(U))
370 'Calculate W(u,r/B) using polynomial approximation
380 L=S/3.75
390 V=1+3.5156229#*L^2+3.0899424#*L^4+1.2067492#*L^6+.265973*L^8+.0360768*L^10
```

```
400 V=V+.0045813*L^12
410 F=S/2
420 H=-LOG(F)*V-.5772156600000009#+.422784*F^2+.23069756#*F^4+.0348859*F^6
430 H=H+2.62698E-03*F^8+.0001075*F^10+.0000074*F^12
440 IF S=0 THEN WS1=WU:GOTO 730
450 N=S^2/(4*U)
460 IF N>5 THEN WS1=2*H:GOTO 730
470 IF U<=.9 THEN 500
480 A=U+.5858:B=U+3.414:C=S*S/4
490 WS1=1.5637*EXP(-A-C/A)/A+4.54*EXP(-B-C/B)/B:GOTO 730
500 IF U<.05 THEN 540
510 IF U>S/2 THEN 530
520 C=-(1.75*U)^-.488*S:WS1=2*H-4.8*10^C:GOTO 730
530 WS1=WU-(S/(4.7*U^.6))^2:GOTO 730
540 IF U>.01 AND S<.1 THEN 530
550 IF N<1 THEN 620
560 WN=N^4+8.573328740100025#*N^3+18.059016973#*N^2+8.634760892499999#*N
570 WN=WN+.2677737343000001#
580 WNP=N^4+9.573322345400009#*N^3+25.6329561486#*N^2+21.0996530827#*N
590 WNP=WNP+3.958496922800012#
600 WN=WN/WNP
610 WN=WN/(N*EXP(N)): GOTO 640
620 WN=-LOG(N)+.9999919300000017#*N-.5772156600000009#-.24991055#*N^2
630 WN=WN+.0551997*N^3-9.76004E-03*N^4+1.07857E-03*N^5
640 WS1=2*H-WN*V:GOTO 730
650 M=-((S-2*U)/(2*U^.5))
660 IF M<0 THEN LET M=ABS(M):GOSUB 680:GOTO 720
670 GOSUB 680:GOTO 710
680 NP=1+.0705230784#*M+.0422820123#*M^2+9.270527200000026D-03*M^3
690 NP=NP+1.52014E-04*M^4+2.76567E-04*M^5+4.30638E-05*M^6
700 N=1/NP^16:RETURN
710 WS1=(3.1416/(2*S))^.5*EXP(-S)*N:GOTO 730
720 WS1=(3.1416/(2*S))^.5*EXP(-S)*(2-N)
730 PRINT:PRINT"Enter Y for partially penetrating wells"
740 INPUT"or N for fully penetrating wells";AP$
750 IF AP$<>"Y" AND AP$<>"y" AND AP$<>"N" AND AP$<>"n" THEN 730
760 IF AP$="N" OR AP$="n" THEN 1110
770 PRINT:PRINT"Program based in part on equations qiven by"
780 PRINT"Hantush,M.S.1964.Hydraulics of wells.In Advances"
790 PRINT"In Hydroscience.Editor,Ven Te Chow.Vol. 1."
800 PRINT"Academic Press,Inc.New York":PRINT
```

```
810 Z=0
820 'Enter partial penetration sub-data base
830 INPUT"AQUIFER HORIZ. HYDR. CONDUCTIVITY (GPD/SQ FT)=";PH
840 LPRINT"AQUIFER HORIZ. HYDR. CONDUCTIVITY (GPD/SQ FT)=" USING AF$;PH
850 INPUT"AQUIFER VERT. HYDR. CONDUCTIVITY (GPD/SQ FT)=";PV
860 LPRINT"AQUIFER VERT. HYDR. CONDUCTIVITY (GPD/SQ FT)=" USING AF$;PV
870 INPUT"RADIAL DISTANCE TO WELL (FT)=";RD
880 LPRINT"RADIAL DISTANCE TO WELL (FT)=" USING AF$;RD
890 INPUT"AQUIFER THICKNESS (FT)=";MA
900 LPRINT"AQUIFER THICKNESS (FT)=" USING AF$;MA
910 INPUT"DISTANCE FROM AQUIFER TOP TO BOTTOM OF PROD. WELL (FT)=";LP
920 LPRINT"DIST. FROM AQUIFER TOP TO BOTTOM OF PROD. WELL (FT)=" USING AF$;LP
930 INPUT"DIST. FROM AQUIFER TOP TO TOP OF PROD.WELL SCREEN (FT)=";DP
940 LPRINT"DIST. FROM AQUIFER TOP TO PROD. WELL SCREEN TOP (FT)" USING AF$;DP
950 INPUT"DISTANCE FROM AQUIFER TOP TO BOTTOM OF OBS. WELL (FT)=";LL
960 LPRINT"DIST. FROM AQUIFER TOP TO BOTTOM OF OBS. WELL (FT)=" USING AF$;LL
970 INPUT"DIST. FROM AQUIFER TOP TO TOP OF OBS. WELL SCREEN (FT)=";DD
980 LPRINT"DIST. FROM AQUIFER TOP TO TOP OF OBS. WELL SCREEN(FT)=" USING AF$;DD
990 LPRINT
1000 'Calculate W(u,r/m(PV/PH)^.5,l/m,d/m,lo/m,do/m) using
1010 'polynomial approximation and summation
1020 FOR K=1 TO 50
1030 CLS:PRINT"COMPUTATIONS ARE IN PROGRESS"
1040 S=K*3.1416*RD*(PV/PH)^.5/MA
1050 GOSUB 1310
1060 GOSUB 1260
1070 NEXT K
1080 WUPAR=2*MA^2*Z/(3.1416^2*(LP-DP)*(LL-DD))
1090 WUPT=WS1+WUPAR
1100 GOTO 1120
1110 WUPT=WS1
1120 CLS
1130 PRINT:PRINT"COMPUTATION RESULTS:"
1140 LPRINT:LPRINT"COMPUTATION RESULTS:"
1150 PRINT:PRINT"WELL FUNCTION=" USING AF$;WUPT
1160 LPRINT:LPRINT"WELL FUNCTION=" USING AF$;WUPT
1170 LPRINT:PRINT
1180 PRINT"Enter Y for another computation"
1190 INPUT"or N to end program";A$
1200 IF A$<>"Y" AND A$<>"y" AND A$<>"N" AND A$<>"n" THEN 1180
1210 IF A$="N" OR A$="n" THEN 1240
```

```
1220 LPRINT:PRINT
1230 GOTO 10
1240 END
1250 'Subroutine to compute sum of sin factors
1260 AA=SIN(K*3.1416*(LP/MA))-SIN(K*3.1416*(DP/MA))
1270 BB=SIN(K*3.1416*(LL/MA))-SIN(K*3.1416*(DD/MA))
1280 Z=Z+1/K^2*(AA*BB*WS)
1290 RETURN
1300 'Subroutine to compute W(u,r/B)
1310 IF U>5! THEN WS=0:GOTO 1790
1320 IF S>2 THEN 1690
1330 IF U>=1 THEN 1360
1340 WU=-LOG(U)-.5772156600000009#+.9999919300000017#*U-.24991055#*U*U
1350 WU=WU+.0551997*U^3-9.76004E-03*U^4+1.07857E-03*U^5:GOTO 1420
1360 WU=U^4+8.573328740100025#*U^3+18.059016973#*U^2+8.634760892499999#*U
1370 WU=WU+.2677737343#
1380 WUP=U^4+9.573322345400009#*U^3+25.6329561486#*U^2+21.0996530827#*U
1390 WUP=WUP+3.958496922800001#
1400 WU=WU/WUP
1410 WU=WU/(U*EXP(U))
1420 L=S/3.75
1430 V=1+3.5156229#*L^2+3.0899424#*L^4+1.2067492#*L^6+.265973*L^8+.0360768*L^10
1440 V=V+.0045813*L^12
1450 F=S/2
1460 H=-LOG(F)*V-.5772156600000009#+.422784*F^2+.23069756#*F^4+.0348859*F^6
1470 H=H+2.62698E-03*F^8+.0001075*F^10+.0000074*F^12
1480 IF S=0 THEN WS=WU:GOTO 1790
1490 N=S^2/(4*U)
1500 IF N>5 THEN WS=2*H:GOTO 1790
1510 IF U<=.9 THEN 1540
1520 A=U+.5858:B=U+3.414:C=S*S/4
1530 WS=1.5637*EXP(-A-C/A)/A+4.54*EXP(-B-C/B)/B:GOTO 1790
1540 IF U<.05 THEN 1580
1550 IF U>S/2 THEN 1570
1560 C=-(1.75*U)^-.488*S:WS=2*H-4.8*10^C:GOTO 1790
1570 WS=WU-(S/(4.7*U^.6))^2:GOTO 1790
1580 IF U>.01 AND S<.1 THEN 1570
1590 IF N<1 THEN 1660
1600 WN=N^4+8.573328740100025#*N^3+18.059016973#*N^2+8.634760892499999#*N
1610 WN=WN+.267773734300001#
1620 WNP=N^4+9.573322345400009#*N^3+25.6329561486#*N^2+21.0996530827#*N
```

```
1630 WNP=WNP+3.958496922800012#
1640 WN=WN/WNP
1650 WN=WN/(N*EXP(N)):GOTO 1680
1660 WN=-LOG(N)+.9999919300000017#*N-.5772156600000009#-.24991055#*N^2
1670 WN=WN+.0551997*N^3-9.76004E-03*N^4+1.07857E-03*N^5
1680 WS=2*H-WN*V:GOTO 1790
1690 M=-((S-2*U)/(2*U^.5))
1700 IF M<0 THEN M=ABS(M):GOSUB 1720:GOTO 1780
1710 GOSUB 1720:GOTO 1770
1720 NP=1+.0705230784#*M+.0422820123#*M^2+9.270527200000026D-03*M^3
1730 NP=NP+1.52014E-04*M^4+2.76567E-04*M^5+4.30638E-05*M^6
1740 IF NP>100! THEN 1760
1750 N=1/NP^16:RETURN
1760 N=0!:RETURN
1770 WS=(3.1416/(2*S))^.5*EXP(-S)*N:GOTO 1790
1780 WS=(3.1416/(2*S))^.5*EXP(-S)*(2-N)
1790 RETURN
```

PT6 SEMILOG TIME-DRAWDOWN OR DISTANCE-DRAWDOWN ANALYSIS

```
10 CLEAR
20 CLS:KEY OFF
30 PRINT"Program: PT6"
40 PRINT"Author : W.C. Walton"
50 PRINT"Version: IBM/PC 2.1 "
60 PRINT"Purpose: CALCULATE AQUIFER TRANSMISSIVITY AND"
70 PRINT"         STORATIVITY FROM SEMILOG TIME- OR DISTANCE-"
80 PRINT"         DRAWDOWN GRAPH FOR PUMPING TEST ANALYSIS"
90 PRINT
100 PRINT"Program based in part on program presented"
110 PRINT"by Poole,L.,M.Borchers,and K.Koessel.1981"
120 PRINT"Some common BASIC programs-TRS-80 Level II"
130 PRINT"Edition.Osborne/McGraw-Hill.pp.145-146.":PRINT
140 PRINT"Press any key to continue":PRINT
150 A$=INKEY$:IF A$="" THEN 150
160 PRINT"For details on equations and assumptions see"
170 PRINT"Walton,W.C.1985.2nd edition.Practical aspects of"
180 PRINT"groundwater modeling.NATIONAL WATER WELL ASSOC."
190 PRINT"Pages 465-470":PRINT
200 DEFDBL A-Z:DEFSNG I
210 AF$="##.####^^^^"
220 PRINT"DATA BASE:":PRINT
230 LPRINT:LPRINT"DATA BASE:":LPRINT
240 PRINT:INPUT"NUMBER OF KNOWN POINTS(MUST BE >5)";N
250 LPRINT"NUMBER OF KNOWN POINTS=";N
260 PRINT:PRINT"Enter T for time-drawdown data or"
270 INPUT"D for distance-drawdown data";TD$
280 IF TD$<>"T" AND TD$<>"t" AND TD$<>"D" AND TD$<>"d" THEN 260
290 PRINT
300 JJ=0
310 KK=0
320 LL=0
330 MM=0
340 R2=0
350 FOR I=1 TO N
360 PRINT"POINT NUMBER=";I
370 LPRINT"POINT NUMBER=";I
380 IF TD$="D" OR TD$="d" THEN 440
```

```
390 INPUT"X (TIME)-COORDINATE OF POINT (MIN)";X
400 INPUT"Y (DRAWDOWN)-COORDINATE OF POINT (FT)=";Y
410 LPRINT"X (TIME)-COORDINATE OF POINT (MIN)=" USING AF$;X
420 LPRINT"Y (DRAWDOWN)-COORDINATE OF POINT (FT)=" USING AF$;Y
430 GOTO 480
440 INPUT"X (DISTANCE)-COORDINATE OF POINT (FT)=";X
450 INPUT"Y (DRAWDOWN)-COORDINATE OF POINT (FT)=";Y
460 LPRINT"X (DISTANCE)-COORDINATE OF POINT (FT)=" USING AF$;X
470 LPRINT"Y (DRAWDOWN)-COORDINATE OF POINT (FT)=" USING AF$;Y
480 X=LOG(X)/LOG(10)
490 'Linear regression
500 JJ=JJ+X
510 KK=KK+Y
520 LL=LL+X^2
530 MM=MM+Y^2
540 R2=R2+X*Y
550 NEXT I
560 BB=(N*R2-KK*JJ)/(N*LL-JJ^2)
570 AA=(KK-BB*JJ)/N
580 PRINT:PRINT"Enter Y to revise data base"
590 INPUT"or N to continue";R$:PRINT
600 IF R$<>"Y" AND R$<>"y" AND R$<>"N" AND R$<>"n" THEN 580
610 IF R$="Y" OR R$="y" THEN 220
620 INPUT"PRODUCTION WELL DISCHARGE RATE (GPM)=";Q
630 LPRINT"PRODUCTION WELL DISCHARGE RATE (GPM)=" USING AF$;Q
640 IF TD$="D" OR TD$="d" THEN 680
650 INPUT"DISTANCE FROM PRODUCTION WELL (FT)=";DIST
660 LPRINT"DISTANCE FROM PRODUCTION WELL (FT)=" USING AF$;DIST
670 GOTO 700
680 INPUT"TIME AFTER PUMPING STARTED (MIN)=";TIME
690 LPRINT"TIME AFTER PUMPING STARTED (MIN)=" USING AF$;TIME
700 X0=2.71828183#^-(AA/(.4342945*BB))
710 IF TD$="D" OR TD$="d" THEN 750
720 TRANS=264*Q/BB
730 STOR=TRANS*X0/(4790*DIST^2)
740 GOTO 770
750 TRANS=264*Q/BB:TRANS=-TRANS
760 STOR=TRANS*TIME/(4790*X0^2)
770 PRINT:PRINT"COMPUTATION RESULTS:":PRINT
780 LPRINT:LPRINT"COMPUTATION RESULTS:":LPRINT
790 PRINT"AQUIFER TRANSMISSIVITY (GPD/FT)=" USING"#######.##";CSNG(TRANS)
```

```
800 LPRINT"AQUIFER TRANSMISSIVITY (GPD/FT)=" USING"#######.##";CSNG(TRANS)
810 PRINT"AQUIFER STORATIVITY (DIM)=" USING"##.###^^^^";CSNG(STOR)
820 LPRINT"AQUIFER STORATIVITY (DIM)=" USING"##.###^^^^";CSNG(STOR)
830 PRINT:PRINT"Enter Y for another calculation or"
840 INPUT"N to end program";E$
850 IF E$<>"Y" AND E$<>"y" AND E$<>"N" AND E$<>"n" THEN 830
860 IF E$="Y" OR E$="y" THEN 10
870 PRINT
880 END
```

PT7 STORATIVITY ANALYSIS NEAR STREAM

```
10 CLS:CLEAR:KEY OFF
20 PRINT
30 PRINT"Program: PT7"
40 PRINT"Author : W.C. Walton"
50 PRINT"Version: IBM/PC 2.1; Copyright 1987 Lewis Publishers, Inc."
60 PRINT"Purpose: DETERMINE STORATIVITY BY METHOD OF"
70 PRINT"         SUCCESSIVE APPROXIMATIONS FOR PUMPING"
80 PRINT"         TEST ANALYSIS"
90 PRINT
100 DEFINT I,J,K
110 LPRINT:LPRINT"DATA BASE:":LPRINT
120 PRINT"Program based in part on equations given by"
130 PRINT"Theis,C.V.1935.The relation between the lowering"
140 PRINT"of piezometric surface and the rate and duration"
150 PRINT"of discharge of a well using ground-water storage."
160 PRINT"Trans. Amer. Geophy. Union.16th Annual Meeting."
170 PRINT:PRINT"Press any key to continue":PRINT
180 A$=INKEY$:IF A$="" THEN 180
190 PRINT"For details on equations and assumptions see"
200 PRINT"Walton,W.C.1985.2nd edition.Practical aspects of"
210 PRINT"groundwater modeling.NATIONAL WATER WELL ASSOC."
220 PRINT"Pages 148,163-170,337-339,394":PRINT
230 PRINT"DATA BASE:":PRINT
240 DIM STOR(24),DD(24)
250 INPUT"AQUIFER TRANSMISSIVITY(GPD/FT)=";TRANS
260 LPRINT"AQUIFER TRANSMISSIVITY(GPD/FT)=" USING"##.##^^^^";TRANS
270 INPUT"TIME AFTER PUMPING STARTED (MIN)=";TIME
280 LPRINT"TIME AFTER PUMPING STARTED (MIN)=" USING"##.###^^^^";TIME
290 INPUT"DISTANCE BETWEEN PROD. AND OBS. WELLS(FT)=";R
300 LPRINT"DIST. BETWEEN PROD. AND OBS. WELLS(FT)=" USING"##.##^^^^";R
310 INPUT"DISTANCE FROM OBS. WELL TO IMAGE WELL (FT)=";RI
320 LPRINT"DISTANCE FROM OBS. WELL TO IMAGE WELL (FT)=" USING "##.###^^^^";RI
330 INPUT"PRODUCTION WELL DISCHARGE (GPM)=";Q
340 LPRINT"PRODUCTION WELL DISCHARGE (GPM)=" USING"######.##";Q
350 INPUT"NUMBER OF STORATIVITY VALUES TO BE TESTED (MUST BE (25)=";TS
360 LPRINT"NUMBER OF STORATIVITY VALUES TO BE TESTED=";TS
370 FOR I=1 TO TS
380 DD(I)=0!
390 PRINT"STORATIVITY TEST NUMBER=";I
```

```
400 LPRINT"STORATIVITY TEST NUMBER="I
410 INPUT"STORATIVITY VALUE TO BE TESTED (DIM)=";STOR(I)
420 LPRINT"STORATIVITY VALUE TO BE TESTED (DIM)=" USING "##.####^^^^";STOR(I)
430 NEXT I
440 PRINT:PRINT"Enter Y to revise data base"
450 INPUT"or N to continue";R$:PRINT
460 IF R$<>"Y" AND R$<>"y" AND R$<>"N" AND R$<>"n" THEN 440
470 IF R$="Y" OR R$="y" THEN 250
480 FOR I=1 TO TS
490 'Calculate u
500 U1=2693*RI^2*STOR(I)/(TRANS*TIME)
510 U=U1:CLS:PRINT"COMPUTATIONS ARE IN PROGRESS":GOSUB 770
520 W1=W
530 U2=2693*R^2*STOR(I)/(TRANS*TIME)
540 U=U2:CLS:PRINT"COMPUTATIONS ARE IN PROGRESS":GOSUB 770
550 W2=W
560 DD(I)=114.6*Q*W2/TRANS
570 DD(I)=DD(I)-114.6*Q*W1/TRANS
580 NEXT I
590 CLS:PRINT:LPRINT
600 PRINT"COMPUTATION RESULTS:":PRINT
610 LPRINT"COMPUTATION RESULTS:":LPRINT
620 PRINT:LPRINT
630 PRINT"STORATIVITY-DRAWDOWN TABLE"
640 LPRINT"STORATIVITY-DRAWDOWN TABLE"
650 PRINT:LPRINT
660 PRINT"STORATIVITY(DIM)          DRAWDOWN(FT)"
670 LPRINT"STORATIVITY(DIM)          DRAWDOWN(FT)"
680 FOR I=1 TO TS
690 PRINT USING"##.#####^^^^";STOR(I);
700 PRINT USING"         #########.##";DD(I)
710 LPRINT USING"##.#####^^^^";STOR(I);
720 LPRINT USING"        #########.##";DD(I)
730 NEXT I
740 PRINT:PRINT
750 END
760 'Subroutine to calculate W(u) using polynomial approximations
770 A=U^2:B=U^3:C=U^4:D=U^5
780 IF U>1 THEN GOTO 810
790 W=-LOG(U)-.577215660000006#+.999991930000007#*U-.24991055#*A+.0551997*B-9
76004E-03*C+1.07857E-03*D
800 GOTO 840
```

```
810 W=C+8.573328740100012#*B+18.059016973#*A+8.634760892499999#*U+.2677737343000
006#
820 W=W/(C+9.573322345400008#*B+25.6329561486#*A+21.0996530827#*U+3.958496922800
006#)
830 W=W/(U*EXP(U))
840 RETURN
```

PT8 STREAM DEPLETION ANALYSIS

```
10 CLEAR
20 CLS:KEY OFF
30 PRINT"Program: PT8"
40 PRINT"Author : W.C. Walton"
50 PRINT"Version: IBM/PC 2.1; Copyright 1987 Lewis Publishers, Inc."
60 PRINT"Purpose: CALCULATE STREAM DEPLETION RATE"
70 PRINT"         FOR PUMPING TEST ANALYSIS"
80 PRINT
90 AF$="##.####^^^^"
100 LPRINT
110 PRINT"Program based in part on equations given by"
120 PRINT"Theis,C.V.1941.The effect of a well on the"
130 PRINT"flow of a nearby stream.Am. Geophys.Union."
140 PRINT"Pt.1":PRINT
150 PRINT"For details on equations and assumptions see"
160 PRINT"Walton,W.C.1985.2nd edition.Practical aspects of"
170 PRINT"groundwater modeling.NATIONAL WATER WELL ASSOC."
180 PRINT"Pages 275-280":PRINT
190 PRINT"DATA BASE:":PRINT
200 LPRINT"DATA BASE:":LPRINT
210 INPUT"PRODUCTION WELL DISCHARGE RATE (GPM)=";A
220 LPRINT"PRODUCTION WELL DISCHARGE RATE (GPM)=" USING AF$;A
230 INPUT"DISTANCE FROM RECHARGING IMAGE WELL TO PROD. WELL (FT)=";B
240 LPRINT"DISTANCE FROM RECHARGING IMAGE WELL TO PROD. WELL (FT)=" USING AF$;B
250 B=B/2
260 INPUT"TIME AFTER PUMPING STARTED (MIN)=";C
270 LPRINT"TIME AFTER PUMPING STARTED (MIN)=" USING AF$;C
280 C=C/1440
290 INPUT"AQUIFER TRANSMISSIVITY (GPD/FT)=";D
300 LPRINT"AQUIFER TRANSMISSIVITY (GPD/FT)=" USING AF$;D
310 INPUT"AQUIFER STORATIVITY (DIM.)=";E
320 LPRINT"AQUIFER STORATIVITY (DIM.)=" USING AF$;E
330 PRINT:PRINT"Enter Y to revise data base"
340 INPUT"or N to continue";R$:PRINT
350 IF R$<>"Y" AND R$<>"y" AND R$<>"N" AND R$<>"n" THEN 330
360 IF R$="Y" OR R$="y" THEN 190
370 'Solve equations for depletion during pumping period
380 G=B/(.535*C*D/E)^.5
390 I=1+.0705230784##G+.0422820123##G^2+9.270527200000007D-03*G^3
```

```
400 J=(I+1.52014E-04*G^4+2.76567E-04*G^5+4.30638E-05*G^6)^16
410 K=A*(1/J)
420 PRINT:PRINT"COMPUTATION RESULTS:"
430 LPRINT:LPRINT"COMPUTATION RESULTS:":LPRINT
440 LPRINT"STREAM DEPLETION AT TIME DURING PUMPING (GPM)=" USING AF$;K
450 PRINT:PRINT"STREAM DEPLETION AT TIME DURING PUMPING (GPM)=" USING AF$;K
460 END
```

PT9 DRAWDOWN BENEATH STREAMBED

```
10 CLEAR
20 CLS:KEY OFF
30 PRINT"Program: PT9"
40 PRINT"Author : W.C. Walton"
50 PRINT"Version: IBM/PC 2.1; Copyright 1987 Lewis Publishers, Inc."
60 PRINT"Purpose: CALCULATE DRAWDOWN BENEATH A STREAMBED"
70 PRINT"         WITH INDUCED INFILTRATION FOR"
80 PRINT"         PUMPING TEST ANALYSIS"
90 PRINT
100 DEFINT I,J,K
110 LPRINT
120 PRINT"Program based in part on equations given by Theis,C.V."
130 PRINT"1935.The relation between the lowering of piezometric"
140 PRINT"surface and the rate and duration of discharge of a well"
150 PRINT"using ground-water storage.Transactions American Geophysical"
160 PRINT"Union. 16th Annual Meeting":PRINT
170 PRINT"Press any key to continue"
180 A$=INKEY$:IF A$="" THEN 180
190 PRINT:PRINT"For details on equations and assumptions see"
200 PRINT"Walton,W.C.1985.2nd edition.Practical aspects of"
210 PRINT"groundwater modeling.NATIONAL WATER WELL ASSOC."
220 PRINT"Pages 148-149":PRINT
230 DIM XOB(30),YOB(30),DD(30,30),XWELL(2),YWELL(2)
240 PRINT"DATA BASE:":PRINT
250 LPRINT"DATA BASE:":LPRINT
260 INPUT"AQUIFER TRANSMISSIVITY (GPD/FT)=";T
270 INPUT"AQUIFER STORATIVITY (DIM.)=";S
280 PRINT:PRINT"FOR SCREEN DISPLAY OF RESULTS LIMIT NO. OF COLUMNS"
290 PRINT"TO 10 AND NO OF ROWS TO 10.":PRINT
300 INPUT"NO. OF GRID COLUMNS(MUST BE <31)=";NC
310 INPUT"NO. OF GRID ROWS (MUST BE <31)=";NR
320 INPUT"GRID SPACING (FT)=";DELTA
330 FOR I=1 TO 2
340 PRINT"WELL NO.=";I
350 IF I=2 THEN 390
360 INPUT"X-COORDINATE OF PRODUCTION WELL (FT)=";XWELL(I)
370 INPUT"Y-COORDINATE OF PRODUCTION WELL (FT)=";YWELL(I)
380 GOTO 410
390 INPUT"X-COORDINATE OF RECHARGING IMAGE WELL (FT)=";XWELL(I)
```

```
400 INPUT"Y-COORDINATE OF RECHARGING IMAGE WELL (FT)=";YWELL(I)
410 NEXT I
420 INPUT"PRODUCTION WELL DISCHARGE (GPM)=";Q
430 INPUT"TIME (MIN)=";TIME
440 INPUT"WELL RADIUS(FT)=";RAD
450 LPRINT"AQUIFER TRANSMISSIVITY (GPD/FT)=";T
460 LPRINT"AQUIFER STORATIVITY (DIM)=";S
470 LPRINT"NO. OF GRID COLUMNS=";NC:LPRINT"NO. OF GRID ROWS=";NR
480 LPRINT"GRID SPACING (FT)=";DELTA:LPRINT"NO. OF WELLS=2"
490 FOR I=1 TO 2
500 LPRINT"WELL NO.=";I
510 IF I=2 THEN 550
520 LPRINT"X-COORDINATE OF PRODUCTION WELL (FT)=" USING"##.####^^^^";XWELL(I)
530 LPRINT"Y-COORDINATE OF PRODUCTION WELL (FT)=" USING"##.####^^^^";YWELL(I)
540 GOTO 570
550 LPRINT"X-COOR. OF RECHARGING IMAGE WELL (FT)=" USING"##.####^^^^";XWELL(I)
560 LPRINT"Y-COOR. OF RECHARGING IMAGE WELL (FT)=" USING"##.####^^^^";YWELL(I)
570 NEXT I
580 LPRINT"WELL DISCHARGE (GPM)=";Q:LPRINT"TIME(MIN)=";TIME
590 LPRINT"WELL RADIUS(FT)=";RAD
600 PRINT:PRINT"Enter Y to revise data base"
610 INPUT"or N to continue";R$:PRINT
620 IF R$<>"Y" AND R$<>"y" AND R$<>"N" AND R$<>"n" THEN 600
630 IF R$="Y" OR R$="y" THEN 240
640 PRINT:LPRINT
650 'Calculate coordinates of grid nodes
660 FOR I=1 TO NC
670 CLS:PRINT"COMPUTATIONS ARE IN PROGRESS"
680 FOR J=1 TO NR
690 XOB(I)=DELTA*I
700 YOB(J)=DELTA*J
710 DD(I,J)=0!
720 NEXT J,I
730 'Calculate distances between wells and grid nodes
740 FOR K=1 TO NC
750 CLS:PRINT"COMPUTATIONS ARE IN PROGRESS"
760 FOR J=1 TO NR
770 FOR I=1 TO 2
780 R2=ABS(XOB(K)-XWELL(I))^2+ABS(YOB(J)-YWELL(I))^2
790 IF R2=0 THEN R2=RAD
800 'Calculate u
```

```
810 U=2693*R2*S/(T*TIME)
820 IF U>5! THEN WU=0!
830 IF U>5! THEN 900
840 IF U>1 THEN GOTO 880
850 'Calculate W(u) with polynomial approximations
860 WU=-LOG(U)-.5772+U-.25*U^2+.0552*U^3-.00976*U^4+.001*U^5
870 GOTO 900
880 WU=((U^4+8.573301*U^3+18.059*U^2+8.63476*U+.267774)/(U^4+9.5773*U^3+25.633*
^2+21.1*U+3.9585))/(U*EXP(U))
890 'Calculate drawdowns at grid nodes
900 IF I=2 THEN DD(K,J)=DD(K,J)-114.6*Q*WU/T
910 IF I=1 THEN DD(K,J)=DD(K,J)+114.6*Q*WU/T
920 NEXT I,J,K
930 CLS:PRINT:PRINT"COMPUTATION RESULTS:"
940 LPRINT"COMPUTATION RESULTS:"
950 CLS:A$="####.##"
960 PRINT:LPRINT
970 IF NC>10 OR NR>10 THEN 1000
980 PRINT"VALUES OF DRAWDOWN (FT) AT NODES:"
990 GOTO 1030
1000 PRINT"Values of drawdown will"
1010 PRINT"not appear on the screen"
1020 PRINT"but will be printed on paper."
1030 LPRINT"VALUES OF DRAWDOWN (FT) AT NODES:"
1040 PRINT:LPRINT
1050 IF NC>10 OR NR>10 THEN 1090
1060 PRINT"J-ROW " SPC(26) "I-COLUMN"
1070 PRINT"        1      2      3      4      5";
1080 PRINT"     6      7      8      9     10":PRINT
1090 LPRINT"J-ROW " SPC(26) "I-COLUMN"
1100 LPRINT"        1      2      3      4      5";
1110 LPRINT"     6      7      8      9     10":LPRINT
1120 IF NC>10 OR NR>10 THEN 1290
1130 IF NC=10 THEN NCC=10
1140 IF NC<10 THEN NCC=NC
1150 IF NR=10 THEN NRR=10
1160 IF NR<10 THEN NRR=NR
1170 FOR J=1 TO NRR
1180 IF J=10 THEN 1210
1190 PRINT" ";J;
1200 GOTO 1220
1210 PRINT"";J;
```

```
1220 FOR I=1 TO NCC
1230 IF I=NCC THEN PRINT USING A$;DD(I,J)
1240 IF I=NCC THEN 1260
1250 PRINT USING A$;DD(I,J);
1260 NEXT I,J:PRINT
1270 PRINT"Press any key to continue"
1280 T$=INKEY$:IF T$="" THEN 1280
1290 IF NC<10 THEN NCC=NC
1300 IF NC=10 OR NC>10 THEN NCC=10
1310 FOR J=1 TO NR
1320 IF J>9 THEN 1350
1330 LPRINT" ";J;
1340 GOTO 1360
1350 LPRINT"";J;
1360 FOR I=1 TO NCC
1370 IF I=NCC THEN LPRINT USING A$;DD(I,J)
1380 IF I=NCC THEN 1400
1390 LPRINT USING A$;DD(I,J);
1400 NEXT I,J:LPRINT:CLS
1410 IF NC<11 THEN 1730
1420 LPRINT"J-ROW " SPC(26) "I-COLUMN"
1430 LPRINT"        11      12      13      14      15";
1440 LPRINT"     16      17      18      19      20":LPRINT
1450 IF NC<20 THEN NCC=NC
1460 IF NC=20 OR NC>20 THEN NCC=20
1470 FOR J=1 TO NR
1480 IF J>9 THEN 1510
1490 LPRINT" ";J;
1500 GOTO 1520
1510 LPRINT"";J;
1520 FOR I=11 TO NCC
1530 IF I=NCC THEN LPRINT USING A$;DD(I,J)
1540 IF I=NCC THEN 1560
1550 LPRINT USING A$;DD(I,J);
1560 NEXT I,J:LPRINT:CLS
1570 IF NC<21 THEN 1730
1580 LPRINT"J-ROW " SPC(26) "I-COLUMN"
1590 LPRINT"        21      22      23      24      25";
1600 LPRINT"     26      27      28      29      30":LPRINT
1610 IF NC<30 THEN NCC=NC
1620 IF NC=30 THEN NCC=30
```

```
1630 FOR J=1 TO NR
1640 IF J>9 THEN 1670
1650 LPRINT" ";J;
1660 GOTO 1680
1670 LPRINT"";J;
1680 FOR I=21 TO NCC
1690 IF I=NCC THEN LPRINT USING A$;DD(I,J)
1700 IF I=NCC THEN 1720
1710 LPRINT USING A$;DD(I,J);
1720 NEXT I,J:LPRINT
1730 PRINT:PRINT
1740 END
```

PT10 WELL LOSS COEFFICIENT

```
10 CLEAR
20 CLS:KEY OFF
30 PRINT"Program: PT10"
40 PRINT"Author : W.C. Walton"
50 PRINT"Version: IBM/PC 2.1; Copyright 1987 Lewis Publishers, Inc."
60 PRINT"Purpose: CALCULATE WELL LOSS COEFFICIENT FOR"
70 PRINT"         PUMPING TEST ANALYSIS"
80 PRINT
90 AF$="##.####^^^^"
100 LPRINT
110 PRINT"Program based in part on equations given by"
120 PRINT"Jacob,C.E.1946.Drawdown test to determine"
130 PRINT"effective radius of artesian well.Proc. Am."
140 PRINT"Soc. Civil Engrs.Vol.79,No.5":PRINT
150 PRINT"For details on equations and assumptions see"
160 PRINT"Walton,W.C.1985.2nd edition.Practical aspects of"
170 PRINT"groundwater modeling.NATIONAL WATER WELL ASOC."
180 PRINT"Pages 128-129.":PRINT
190 PRINT"DATA BASE:":PRINT
200 LPRINT"DATA BASE:":LPRINT
210 INPUT"PUMPING RATE-1 (GPM)=";A
220 LPRINT"PUMPING RATE-1 (GPM)=" USING AF$;A
230 INPUT"PUMPING RATE-2 (GPM)=";B
240 LPRINT"PUMPING RATE-2 (GPM)=" USING AF$;B
250 INPUT"PUMPING RATE-3 (GPM)=";C
260 LPRINT"PUMPING RATE-3 (GPM)=" USING AF$;C
270 INPUT"DRAWDOWN CHANGE-1 (FT)=";D
280 LPRINT"DRAWDOWN CHANGE-1 (FT)=" USING AF$;D
290 INPUT"DRAWDOWN CHANGE-2 (FT)=";E
300 LPRINT"DRAWDOWN CHANGE-2 (FT)=" USING AF$;E
310 INPUT"DRAWDOWN CHANGE-3 (FT)=";F
320 LPRINT"DRAWDOWN CHANGE-3 (FT)=" USING AF$;F
330 PRINT:PRINT"Enter Y to revise data base"
340 INPUT"or N to continue";R$:PRINT
350 IF R$<>"Y" AND R$<>"y" AND R$<>"N" AND R$<>"n" THEN 330
360 'Solve well loss coefficient equations
370 IF R$="Y" OR R$="y" THEN 190
380 G=A/448.8
390 H=(B-A)/448.8
```

```
400 I=(C-B)/448.8
410 PRINT"COMPUTATION RESULTS:"
420 LPRINT:LPRINT"COMPUTATION RESULTS:"
430 J=((E/H)-(D/G))/(G+H)
440 LPRINT:LPRINT"WELL LOSS CONSTANT 1-2 (SEC^2/FT^5)=";J
450 PRINT:PRINT"WELL LOSS CONSTANT 1-2 (SEC^2/FT^5):";J
460 K=((F/I)-(E/H))/(H+I)
470 LPRINT"WELL LOSS CONSTANT 2-3 (SEC^2/FT^5)=";K
480 PRINT"WELL LOSS CONSTANT 2-3 (SEC^2/FT^5):";K
490 L=((F/I)-((D+E)/(G+H)))/(G+H+I)
500 LPRINT"WELL LOSS CONSTANT (1+2)-3 (SEC^2/FT^5)=";L
510 PRINT"WELL LOSS CONSTANT (1+2)-3 (SEC^2/FT^5):";L
520 M=((E+F)/(H+I)-(D/G))/(G+H+I)
530 LPRINT"WELL LOSS CONSTANT 1-(2+3)(SEC^2/FT^5)=";M
540 PRINT"WELL LOSS CONSTANT 1-(2+3) (SEC^2/FT^5):";M
550 PRINT:PRINT"Enter Y for another computation"
560 INPUT"or N to end program";CO$
570 IF CO$<>"Y" AND CO$<>"y" AND CO$<>"N" AND CO$<>"n" THEN 550
580 IF CO$="Y" OR CO$="y" THEN 10
590 END
```

PT11 SOLUTION OF EQUATIONS IN TEXT

```
10 CLS:CLEAR:KEY OFF
20 PRINT"Program: PT11"
30 PRINT"Author : W.C. Walton"
40 PRINT"Version:IBM/PC 2.1; Copyright 1987 Lewis Publishers, Inc."
50 PRINT"Purpose: SOLVE PUMPING TEST DESIGN AND"
60 PRINT"          ANALYSIS EQUATIONS"
70 PRINT:PRINT"Enter number of equation to be solved"
80 PRINT"If equation no. is 2.10 then enter 2.01"
90 PRINT"If equation no. is 3.10 then enter 3.01"
100 PRINT"If equation no. is 3.20 then enter 3.02"
110 PRINT"If equation no. is 3.30 then enter 3.03"
120 INPUT"If equation no. is 3.40 then enter 3.04:";NO
130 IF NO=1.1 THEN 850
140 IF NO=1.2 THEN 940
150 IF NO=1.3 THEN 1020
160 IF NO=1.4 THEN 1100
170 IF NO>1.4 AND NO<2.01 THEN 70
180 IF NO=2.1 THEN 1190
190 IF NO=2.2 THEN 1270
200 IF NO=2.3 THEN 1340
210 IF NO=2.4 THEN 1380
220 IF NO=2.5 THEN 1420
230 IF NO=2.6 THEN 1540
240 IF NO=2.7 THEN 1610
250 IF NO=2.8 THEN 1700
260 IF NO=2.9 THEN 1750
270 IF NO=2.01 THEN 1800
280 IF NO=2.11 THEN 1850
290 IF NO=2.12 THEN 1900
300 IF NO=2.13 THEN 1950
310 IF NO=2.14 THEN 2000
320 IF NO>2.14 AND NO<3.01 THEN 70
330 IF NO=3.1 THEN 2050
340 IF NO=3.2 THEN 2110
350 IF NO=3.3 THEN 2180
360 IF NO=3.4 THEN 2240
370 IF NO=3.5 THEN 2290
380 IF NO=3.6 THEN 2350
390 IF NO=3.7 THEN 2410
```

```
400 IF NO=3.8 THEN 2470
410 IF NO=3.9 THEN 2560
420 IF NO=3.01 THEN 2620
430 IF NO=3.11 THEN 2690
440 IF NO=3.12 THEN 2760
450 IF NO=3.13 THEN 2830
460 IF NO=3.14 THEN 2880
470 IF NO=3.15 THEN 2930
480 IF NO=3.16 THEN 2990
490 IF NO=3.17 THEN 3060
500 IF NO=3.18 THEN 3110
510 IF NO=3.19 THEN 3160
520 IF NO=3.02 THEN 3220
530 IF NO=3.21 THEN 3280
540 IF NO=3.22 THEN 3350
550 IF NO=3.23 THEN 3420
560 IF NO=3.24 THEN 3480
570 IF NO=3.25 THEN 3550
580 IF NO=3.26 THEN 3610
590 IF NO=3.27 THEN 3680
600 IF NO=3.28 THEN 3760
610 IF NO=3.29 THEN 3820
620 IF NO=3.03 THEN 3890
630 IF NO=3.31 THEN 3990
640 IF NO=3.32 THEN 4100
650 IF NO=3.33 THEN 4160
660 IF NO=3.34 THEN 4230
670 IF NO=3.35 THEN 4300
680 IF NO=3.36 THEN 4360
690 IF NO=3.37 THEN 4420
700 IF NO=3.38 THEN 4490
710 IF NO=3.39 THEN 4560
720 IF NO=3.04 THEN 4610
730 IF NO=3.41 THEN 4670
740 IF NO=3.42 THEN 4720
750 IF NO=3.43 THEN 4780
760 IF NO=3.44 THEN 4860
770 IF NO=3.45 THEN 4920
780 IF NO=3.46 THEN 4980
790 IF NO>3.46 AND NO<4.01 THEN 70
800 IF NO=4.1 THEN 5080
```

```
810 IF NO=4.2 THEN 5130
820 IF NO=4.3 THEN 5210
830 IF NO )4.3 THEN 70
840 GOTO 70
850 PRINT:INPUT"PRODUCTION WELL EFFECTIVE RADIUS (FT)=";RW
860 INPUT"PUMP-COLUMN PIPE RADIUS (FT)=";RC
870 INPUT"AQUIFER TRANSMISSIVITY (GPD/FT)=";TRANS
880 TW=540000!*(RW^2-RC^2)/TRANS
890 PRINT:PRINT"TIME AFTER PUMPING STARTED BEYOND"
900 PRINT"WHICH WELL STORAGE CAPACITY IMPACTS ARE"
910 PRINT"NEGLIGIBLE (LESS THAN 1 PERCENT OF "
920 PRINT"DRAWDOWN VALUES) (MIN)=" USING"#######.##";TW
930 GOTO 5280
940 PRINT:INPUT"AQUIFER THICKNESS (FT)=";THICK
950 INPUT"AQUIFER HORIZONTAL HYDRAULIC CONDUCTIVITY (GPD/SQ FT)=";PH
960 INPUT"AQUIFER VERTICAL HYDRAULIC CONDUCTIVITY (GPD/SQ FT)=";PV
970 RP=1.5*THICK*(PH/PV)^.5
980 PRINT:PRINT"DISTANCE FROM PRODUCTION WELL BEYOND"
990 PRINT"WHICH PARTIAL PENETRATION IMPACTS ARE"
1000 PRINT"NEGLIGIBLE (FT)=" USING"#######.##";RP
1010 GOTO 5280
1020 PRINT:INPUT"AQUIFER THICKNESS (FT)=";THICK
1030 INPUT"AQUIFER WATER TABLE STORATIVITY (DIM)=";STOR
1040 INPUT"AQUIFER VERTICAL HYDRAULIC CONDUCTIVITY (GPD/SQ FT)=";PV
1050 TD=54000!*THICK*STOR/PV
1060 PRINT:PRINT"TIME AFTER PUMPING STARTED BEYOND"
1070 PRINT"WHICH DELAYED GRAVITY YIELD IMPACTS ARE"
1080 PRINT"NEGLIGIBLE (MIN)=" USING"#######.##";TD
1090 GOTO 5280
1100 PRINT:INPUT"AQUIFER TRANSMISSIVITY (GPD/FT)=";TRANS
1110 INPUT"TIME AFTER PUMPING STARTED (MIN)=";TIME
1120 INPUT"ARTESIAN OR WATER TABLE STORATIVITY (DIM)=";STOR
1130 RN=.043*(TRANS*TIME/STOR)^.5
1140 PRINT:PRINT"DISTANCE FROM OBSERVATION WELL BEYOND"
1150 PRINT"WHICH BOUNDARY AND DISCONTINUITY IMAGE WELL"
1160 PRINT"OR NEARBY PRODUCTION WELL IMPACTS"
1170 PRINT"ARE NEGLIGIBLE (FT)=" USING"#######.##";RN
1180 GOTO 5280
1190 PRINT:PRINT"DISTANCE FROM OBSERVATION WELL TO"
1200 INPUT"BOUNDARY IMAGE WELL (FT)=";DI
1210 INPUT"AQUIFER TRANSMISSIVITY (GPD/FT)=";TRANS
```

```
1220 INPUT"AQUIFER STORATIVITY (DIM)=";STOR
1230 TI=5400!*DI^2*STOR/TRANS
1240 PRINT:PRINT"TIME PUMPING TEST DURATION MUST EXCEED"
1250 PRINT"BEFORE IMAGE WELL IMPACTS ARE CLEAR (MIN)=" USING"#######.##";TI
1260 GOTO 5280
1270 PRINT:INPUT"PRODUCTION WELL DISCHARGE RATE (GPM)=";Q
1280 INPUT"AQUIFER TRANSMISSIVITY AT PRODUCTION WELL (GPD/FT)=";TP
1290 INPUT"AQUIFER TRANSMISSIVITY BEYOND DISCONTINUITY (GPD/FT)=";TD
1300 DT=(TP-TD)/(TP+TD)
1310 QI=Q*DT
1320 PRINT:PRINT"IMAGE WELL STRENGTH (GPM)=" USING"######.##";QI
1330 GOTO 5280
1340 PRINT:INPUT"ARTESIAN AQUIFER TRANSMISSIVITY (GPD/FT)=";TRANS
1350 QS=TRANS/2000
1360 PRINT:PRINT"ARTESIAN PROD. WELL SPEC. CAP. (GPM/FT)=" USING"#####.##";QS
1370 GOTO 5280
1380 PRINT:INPUT"WATER TABLE AQUIFER TRANSMISSIVITY (GPD/FT)=";TRANS
1390 QS=TRANS/1500
1400 PRINT:PRINT"PROD. WELL SPECIFIC CAPACITY (GPM/FT)=" USING"#####.##";QS
1410 GOTO 5280
1420 PRINT:PRINT"PRODUCTION WELL SPECIFIC CAPACITY WITH"
1430 INPUT"FULL PENETRATION (GPM/FT)=";QF
1440 INPUT"PRODUCTION WELL EFFECTIVE RADIUS (FT)=";RD
1450 INPUT"PRODUCTION WELL SCREEN LENGTH (FT)=";LS
1460 INPUT"AQUIFER THICKNESS (FT)=";THICK
1470 L=LS/THICK
1480 M=RD/(2*THICK*L)
1490 N=COS(3.1416*L/2)
1500 QP=QF*L*(1+7*(M*N)^.5)
1510 PRINT:PRINT"PRODUCTION WELL SPECIFIC CAPACITY WITH"
1520 PRINT"PARTIAL PENETRATION (GPM/FT)=" USING"######.##";QP
1530 GOTO 5280
1540 PRINT:INPUT"DISTANCE FROM PRODUCTION WELL (FT)=";R
1550 INPUT"AQUIFER STORATIVITY (DIM)=";STOR
1560 INPUT"AQUIFER TRANSMISSIVITY (GPD/FT)=";TRANS
1570 INPUT"TIME AFTER PUMPING STARTED (MIN)=";TIME
1580 U=2693*R^2*STOR/(TRANS*TIME)
1590 PRINT:PRINT"AQUIFER U=" USING"##.####^^^^";U
1600 GOTO 5280
1610 PRINT:PRINT"DISTANCE ABOVE AQUIFER TOP TO"
1620 INPUT"BOTTOM OF AQUITARD OBS. WELL (FT)=";Z
```

```
1630 INPUT"AQUITARD STORATIVITY (DIM)=";STORC
1640 INPUT"AQUITARD VERTICAL HYDRAULIC CONDUCTIVITY (GPD/SQ FT)=";PC
1650 INPUT"AQUITARD THICKNESS (FT)=";THICK
1660 INPUT"TIME AFTER PUMPING STARTED (MIN)=";TIME
1670 UC=2693*Z^2*STORC/(PC*THICK*TIME)
1680 PRINT:PRINT"AQUITARD U=" USING"##.####^^^^";UC
1690 GOTO 5280
1700 PRINT:INPUT"CHANGE IN WATER LEVEL (FT)=";W
1710 INPUT"CHANGE IN BAROMETRIC PRESSURE (FT OF WATER)=";B
1720 BE=(W/B)*100
1730 PRINT:PRINT"BAROMETRIC EFFICIENCY (PERCENT)=" USING"####.##";BE
1740 GOTO 5280
1750 PRINT:INPUT"BAROMETRIC EFFICIENCY (PERCENT)=";BE
1760 INPUT"CHANGE IN BAROMETRIC PRESSURE (FEET OF WATER)=";B
1770 W=(BE*B)/100
1780 PRINT:PRINT"CHANGE IN WATER LEVEL (FT)=" USING"####.###";W
1790 GOTO 5280
1800 PRINT:INPUT"CHANGE IN WATER LEVEL (FT)=";WS
1810 INPUT"CHANGE IN SURFACE STAGE (FT)=";H
1820 TE=(WS/H)*100
1830 PRINT:PRINT"TIDAL EFFICIENCY (PERCENT)=" USING"####.##";TE
1840 GOTO 5280
1850 PRINT:INPUT"CHANGE IN WATER LEVEL (FT)=";WS
1860 INPUT"CHANGE IN SURFACE STAGE (FT)=";H
1870 RE=(WS/H)*100
1880 PRINT:PRINT"RIVER EFFICIENCY (PERCENT)=" USING"####.##";RE
1890 GOTO 5280
1900 PRINT:INPUT"TIDAL EFFICIENCY (PERCENT)=";TE
1910 INPUT"CHANGE IN SURFACE STAGE (FT)=";H
1920 WS=(TE*H)/100
1930 PRINT:PRINT"CHANGE IN WATER LEVEL (FT)=" USING"####.###";WS
1940 GOTO 5280
1950 PRINT:INPUT"RIVER EFFICIENCY (PERCENT)=";RE
1960 INPUT"CHANGE IN SURFACE STAGE (FT)=";H
1970 WS=(RE*H)/100
1980 PRINT:PRINT"CHANGE IN WATER LEVEL (FT)=" USING"####.###";WS
1990 GOTO 5280
2000 PRINT:INPUT"OBSERVED DRAWDOWN (FT)=";SO
2010 INPUT"AQUIFER THICKNESS (FT)=";M
2020 SA=SO-SO^2/(2*M)
2030 PRINT:PRINT"ADJUSTED DRAWDOWN (FT)=" USING"####.###";SA
```

```
2040 GOTO 5280
2050 PRINT:INPUT"DISCHARGE RATE (GPM)=";Q
2060 INPUT"WELL FUNCTION (DIM)=";WU
2070 INPUT"AQUIFER TRANSMISSIVITY (GPD/FT)=";TRANS
2080 DS=114.6*Q*WU/TRANS
2090 PRINT:PRINT"DRAWDOWN (FT)=" USING"#####.##";DS
2100 GOTO 5280
2110 PRINT:INPUT"DISTANCE FROM PRODUCTION WELL (FT)=";R
2120 INPUT"AQUIFER STORATIVITY (DIM)=";STOR
2130 INPUT"AQUIFER TRANSMISSIVITY (GPD/FT)=";TRANS
2140 INPUT"TIME AFTER PUMPING STARTED (MIN)=";TIME
2150 U=2693*R^2*STOR/(TRANS*TIME)
2160 PRINT:PRINT"U=" USING"##.####^^^^";U
2170 GOTO 5280
2180 PRINT:INPUT"DISCHARGE RATE (GPM)=";Q
2190 INPUT"WELL FUNCTION=";WU
2200 INPUT"AQUIFER TRANSMISSIVITY (GPD/FT)=";TRANS
2210 DS=114.6*Q*WU/TRANS
2220 PRINT:PRINT"DRAWDOWN (FT)=" USING"#####.##";DS
2230 GOTO 5280
2240 PRINT:INPUT"DISTANCE FROM PRODUCTION WELL (FT)=";R
2250 INPUT"PRODUCTION WELL EFFECTIVE RADIUS (FT)=";RW
2260 RHO=R/RW
2270 PRINT:PRINT"RHO=" USING"##.####^^^^";RHO
2280 GOTO 5280
2290 PRINT:INPUT"DISCHARGE RATE (GPM)=";Q
2300 INPUT"WELL FUNCTION=";WU
2310 INPUT"AQUIFER TRANSMISSIVITY (GPD/FT)=";TRANS
2320 DS=114.6*Q*WU/TRANS
2330 PRINT:PRINT"DRAWDOWN (FT)=" USING"#####.##";DS
2340 GOTO 5280
2350 PRINT:INPUT"AQUIFER TRANSMISSIVITY (GPD/FT)=";TRANS
2360 INPUT"AQUITARD THICKNESS (FT)=";MC
2370 INPUT"AQUIFER VERTICAL HYDRAULIC CONDUCTIVITY (GPD/SQ FT)=";PC
2380 B=(TRANS*MC/PC)^.5
2390 PRINT:PRINT"B=" USING"##.####^^^^";B
2400 GOTO 5280
2410 PRINT:INPUT"DISCHARGE RATE (GPM)=";Q
2420 INPUT"WELL FUNCTION (DIM)=";WU
2430 INPUT"AQUIFER TRANSMISSIVITY (GPD/FT)=";TRANS
2440 DS=114.6*Q*WU/TRANS
```

```
2450 PRINT:PRINT"DRAWDOWN (FT)=" USING"#####.##";DS
2460 GOTO 5280
2470 PRINT:INPUT"DISTANCE FROM PRODUCTION WELL (FT)=";R
2480 INPUT"AQUITARD STORATIVITY (DIM)=";SC
2490 INPUT"AQUITARD VERTICAL HYDRAULIC CONDUCTIVITY (GPD/SQ FT)=";PC
2500 INPUT"AQUIFER TRANSMISSIVITY (GPD/FT)=";TRANS
2510 INPUT"AQUIFER STORATIVITY (DIM)=";STOR
2520 INPUT"AQUITARD THICKNESS (FT)=";MC
2530 GAMMA=(R/4)*(SC*PC/(TRANS*STOR*MC))^.5
2540 PRINT:PRINT"GAMMA=" USING"##.####^^^^";GAMMA
2550 GOTO 5280
2560 PRINT:INPUT"DISCHARGE RATE (GPM)=";Q
2570 INPUT"WELL FUNCTION (DIM)=";WU
2580 INPUT"AQUIFER TRANSMISSIVITY (GPD/FT)=";TRANS
2590 DS=114.6*Q*WU/TRANS
2600 PRINT:PRINT"DRAWDOWN (FT)=" USING"#####.##";DS
2610 GOTO 5280
2620 PRINT:INPUT"DISTANCE FROM PRODUCTION WELL (FT)=";R
2630 INPUT"AQUIFER STORATIVITY (DIM)=";STOR
2640 INPUT"AQUIFER TRANSMISSIVITY (GPD/FT)=";TRANS
2650 INPUT"TIME AFTER PUMPING STARTED (MIN)=";TIME
2660 UA=2693*R^2*STOR/(TRANS*TIME)
2670 PRINT:PRINT"UA=" USING"##.####^^^^";UA
2680 GOTO 5280
2690 PRINT:INPUT"DISTANCE FROM PRODUCTION WELL (FT)=";R
2700 INPUT"AQUIFER SPECIFIC YIELD (DIM)=";SY
2710 INPUT"AQUIFER TRANSMISSIVITY (GPD/FT)=";TRANS
2720 INPUT"TIME AFTER PUMPING STARTED (MIN)=";TIME
2730 UB=2693*R^2*SY/(TRANS*TIME)
2740 PRINT:PRINT"UB=" USING"##.####^^^^";UB
2750 GOTO 5280
2760 PRINT:INPUT"DISTANCE FROM PRODUCTION WELL (FT)=";R
2770 INPUT"AQUIFER VERTICAL HYDRAULIC CONDUCTIVITY (GPD/SQ FT)=";PV
2780 INPUT"AQUIFER HORIZONTAL HYDRAULIC CONDUCTIVITY (GPD/SQ FT)=";PH
2790 INPUT"AQUIFER THICKNESS (FT)=";M
2800 BETA=R^2*PV/(M^2*PH)
2810 PRINT:PRINT"BETA=" USING"##.####^^^^";BETA
2820 GOTO 5280
2830 PRINT:INPUT"PRODUCTION WELL FUNCTION (DIM)=";WU
2840 INPUT"IMAGE WELL FUNCTION (DIM)=";WUI
2850 WT=WU+WUI
```

```
2860 PRINT:PRINT"TOTAL WELL FUNCTION (DIM)=" USING"##.####^^^^";WT
2870 GOTO 5280
2880 PRINT:INPUT"DRAWDOWN DUE TO PRODUCTION WELL (FT)=";DS
2890 INPUT"DRAWDOWN DUE TO IMAGE WELL (FT)=";DSI
2900 ST=DS+DSI
2910 PRINT:PRINT"TOTAL DRAWDOWN (FT)=" USING"#####.##";ST
2920 GOTO 5280
2930 PRINT:INPUT"DISCHARGE RATE (GPM)=";Q
2940 INPUT"IMAGE WELL FUNCTION (DIM)=";WU
2950 INPUT"AQUIFER TRANSMISSIVITY (GPD/FT)=";TRANS
2960 DS=114.6*Q*WU/TRANS
2970 PRINT:PRINT"DRAWDOWN (FT)=" USING"#####.##";DS
2980 GOTO 5280
2990 PRINT:INPUT"DISTANCE FROM IMAGE WELL (FT)=";RI
3000 INPUT"AQUIFER STORATIVITY (DIM)=";STOR
3010 INPUT"AQUIFER TRANSMISSIVITY (GPD/FT)=";TRANS
3020 INPUT"TIME AFTER PUMPING STARTED (MIN)=";TIME
3030 UI=2693*RI^2*STOR/(TRANS*TIME)
3040 PRINT:PRINT"UI=" USING"##.####^^^^";UI
3050 GOTO 5280
3060 PRINT:INPUT"FULL PENETRATION WELL FUNCTION (DIM)=";WU
3070 INPUT"PARTIAL PENETRATION WELL FUNCTION (DIM)=";WUP
3080 WT=WU+WUP
3090 PRINT:PRINT"TOTAL WELL FUNCTION (DIM)=" USING"##.####^^^^";WT
3100 GOTO 5280
3110 PRINT:INPUT"FULL PENETRATION DRAWDOWN (FT)=";DS
3120 INPUT"PARTIAL PENETRATION DRAWDOWN (FT)=";SP
3130 ST=DS+SP
3140 PRINT:PRINT"TOTAL DRAWDOWN (FT)=" USING"#####.##";ST
3150 GOTO 5280
3160 PRINT:INPUT"DISCHARGE RATE (GPM)=";Q
3170 INPUT"PARTIAL PENETRATION WELL FUNCTION (DIM)=";WUP
3180 INPUT"AQUIFER TRANSMISSIVITY (GPD/FT)=";TRANS
3190 SP=114.6*Q*WUP/TRANS
3200 PRINT:PRINT"DRAWDOWN DUE TO PARTIAL PENETRATION (FT)=" USING"#####.##";SP
3210 GOTO 5280
3220 PRINT:INPUT"DISCHARGE RATE (GPM)=";Q
3230 INPUT"WELL FUNCTION MATCH POINT COORDINATE (DIM)=";WU
3240 INPUT"DRAWDOWN MATCH POINT COORDINATE (FT)=";SM
3250 TRANS=114.6*Q*WU/SM
3260 PRINT:PRINT"AQUIFER TRANSMISSIVITY (GPD/FT)=" USING"##.####^^^^";TRANS
```

```
3270 GOTO 5280
3280 PRINT:INPUT"AQUIFER TRANSMISSIVITY (GPD/FT)=";TRANS
3290 INPUT"TIME MATCH POINT COORDINATE (MIN)=";TIME
3300 INPUT"DISTANCE FROM PRODUCTION WELL (FT)=";R
3310 INPUT"1/U MATCH POINT COORDINATE (DIM)=";U
3320 STOR=TRANS*TIME/(2693*R^2*U)
3330 PRINT:PRINT"AQUIFER STORATIVITY (DIM)=" USING"##.####^^^^";STOR
3340 GOTO 5280
3350 PRINT:INPUT"AQUIFER TRANSMISSIVITY (GPD/FT)=";TRANS
3360 INPUT"U MATCH POINT COORDINATE (DIM)=";U
3370 INPUT"TIME AFTER PUMPING STARTED (MIN)=";TIME
3380 INPUT"DISTANCE SQUARED MATCH POINT COORDINATE (SQ FT)=";R2
3390 STOR=TRANS*U*TIME/(2693*R2)
3400 PRINT:PRINT"AQUIFER STORATIVITY (DIM)=" USING"##.####^^^^";STOR
3410 GOTO 5280
3420 PRINT:INPUT"DISCHARGE RATE (GPM)=";Q
3430 INPUT"WELL FUNCTION MATCH POINT COORDINATE (DIM)=";WU
3440 INPUT"DRAWDOWN MATCH POINT COORDINATE (DIM)=";SM
3450 TRANS=114.6*Q*WU/SM
3460 PRINT:PRINT"AQUIFER TRANSMISSIVITY (GPD/FT)=" USING"##.####^^^^";TRANS
3470 GOTO 5280
3480 PRINT:INPUT"AQUIFER TRANSMISSIVITY (GPD/FT)=";TRANS
3490 INPUT"TIME MATCH POINT COORDINATE (MIN)=";TIME
3500 INPUT"DISTANCE FROM PRODUCTION WELL (FT)=";R
3510 INPUT"1/U MATCH POINT COORDINATE (DIM)=";U
3520 STOR=TRANS*TIME/(2693*R^2*U)
3530 PRINT:PRINT"AQUIFER STORATIVITY (DIM)=" USING"##.####^^^^";STOR
3540 GOTO 5280
3550 PRINT:INPUT"DISCHARGE RATE (GPM)=";Q
3560 INPUT"WELL FUNCTION MATCH POINT COORDINATE (DIM)=";WU
3570 INPUT"DRAWDOWN MATCH POINT COORDINATE (FT)=";DS
3580 TRANS=114.6*Q*WU/DS
3590 PRINT:PRINT"AQUIFER TRANSMISSIVITY (GPD/FT)=" USING"##.####^^^^";TRANS
3600 GOTO 5280
3610 PRINT:INPUT"AQUIFER TRANSMISSIVITY (GPD/FT)=";TRANS
3620 INPUT"TIME MATCH POINT COORDINATE (MIN)=";TIME
3630 INPUT"DISTANCE FROM PRODUCTION WELL (FT)=";R
3640 INPUT"1/U MATCH POINT COORDINATE (DIM)=";UM
3650 STOR=TRANS*TIME/(2693*R^2*UM)
3660 PRINT:PRINT"AQUIFER STORATIVITY (DIM)=" USING"##.####^^^^";STOR
3670 GOTO 5280
```

```
3680 PRINT:INPUT"AQUIFER TRANSMISSIVITY (GPD/FT)=";TRANS
3690 INPUT"AQUITARD THICKNESS (FT)=";MC
3700 INPUT"R/B MATCH POINT COORDINATE (DIM)=";RB
3710 INPUT"DISTANCE FROM PRODUCTION WELL (FT)=";R
3720 PC=TRANS*MC*RB^2/R^2
3730 PRINT:PRINT"AQUITARD VERTICAL HYDRAULIC"
3740 PRINT"CONDUCTIVITY (GPD/SQ FT)=" USING"##.####^^^^";PC
3750 GOTO 5280
3760 PRINT:INPUT"DISCHARGE RATE (GPM)=";Q
3770 INPUT"WELL FUNCTION MATCH POINT COORDINATE (DIM)=";WU
3780 INPUT"DRAWDOWN MATCH POINT COORDINATE (FT)=";DS
3790 TRANS=114.6*Q*WU/DS
3800 PRINT:PRINT"AQUIFER TRANSMISSIVITY (GPD/FT)=" USING"##.####^^^^";TRANS
3810 GOTO 5280
3820 PRINT:INPUT"AQUIFER TRANSMISSIVITY (GPD/FT)=";TRANS
3830 INPUT"TIME MATCH POINT COORDINATE (MIN)=";TIME
3840 INPUT"DISTANCE FROM PRODUCTION WELL (FT)=";R
3850 INPUT"1/U MATCH POINT COORDINATE (DIM)=";U
3860 STOR=TRANS*TIME/(2693*R^2*U)
3870 PRINT:PRINT"AQUIFER STORATIVITY (DIM)=" USING"##.####^^^^";STOR
3880 GOTO 5280
3890 PRINT:INPUT"GAMMA MATCH POINT COORDINATE (DIM)=";GAMMA
3900 INPUT"AQUIFER TRANSMISSIVITY (GPD/FT)=";TRANS
3910 INPUT"AQUIFER STORATIVITY (DIM)=";STOR
3920 INPUT"AQUITARD THICKNESS (FT)=";MC
3930 INPUT"AQUITARD STORATIVITY (DIM)=";SC
3940 INPUT"DISTANCE FROM PRODUCTION WELL (FT)=";R
3950 PC=GAMMA^2*16*TRANS*STOR*MC/(R^2*SC)
3960 PRINT:PRINT"AQUITARD VERTICAL HYDRAULIC"
3970 PRINT"CONDUCTIVITY (GPD/SQ FT)=" USING"##.####^^^^";PC
3980 GOTO 5280
3990 PRINT:PRINT"VERTICAL DISTANCE FROM AQUIFER TOP TO"
4000 INPUT"BASE OF AQUITARD OBSERVATION WELL (FT)=";Z
4010 PRINT"1/UC VALUE CORRESPONDING TO OBSERVED VALUE"
4020 INPUT"OF SC/S AND CALCULATED VALUE OF U (DIM)=";UC
4030 INPUT"AQUITARD STORATIVITY (DIM)=";SC
4040 INPUT"TIME AFTER PUMPING STARTED (MIN)=";TIME
4050 INPUT"AQUITARD THICKNESS (FT)=";MC
4060 PC=2693*Z^2*UC*SC/(TIME*MC)
4070 PRINT:PRINT"AQUITARD VERTICAL HYDRAULIC"
4080 PRINT"CONDUCTIVITY (GPD/SQ FT)=" USING"##.####^^^^";PC
```

```
4090 GOTO 5280
4100 PRINT:INPUT"DISCHARGE RATE (GPM)=";Q
4110 INPUT"WELL FUNCTION MATCH POINT COORDINATE (DIM)=";WU
4120 INPUT"DRAWDOWN MATCH POINT COORDINATE (FT)=";DS
4130 TRANS=114.6*Q*WU/DS
4140 PRINT:PRINT"AQUIFER TRANSMISSIVITY (GPD/FT)=" USING"##.####^^^^";TRANS
4150 GOTO 5280
4160 PRINT:INPUT"AQUIFER TRANSMISSIVITY (GPD/FT)=";TRANS
4170 INPUT"TIME MATCH POINT COORDINATE (MIN)=";TIME
4180 INPUT"DISTANCE FROM PRODUCTION WELL (FT)=";R
4190 INPUT"1/UA MATCH POINT COORDINATE (DIM)=";UA
4200 STOR=TRANS*TIME/(2693*R^2*UA)
4210 PRINT:PRINT"AQUIFER STORATIVITY (DIM)=" USING"##.####^^^^";STOR
4220 GOTO 5280
4230 PRINT:INPUT"AQUIFER TRANSMISSIVITY (GPD/FT)=";TRANS
4240 INPUT"TIME MATCH POINT COORDINATE (MIN)=";TIME
4250 INPUT"DISTANCE FROM PRODUCTION WELL (FT)=";R
4260 INPUT"1/UB MATCH POINT COORDINATE (DIM)=";UB
4270 STOR=TRANS*TIME/(2693*R^2*UB)
4280 PRINT:PRINT"AQUIFER SPECIFIC YIELD (DIM)=" USING"##.####^^^^";STOR
4290 GOTO 5280
4300 PRINT:INPUT"AQUIFER THICKNESS (FT)=";M
4310 INPUT"BETA MATCH POINT COORDINATE (DIM)=";BETA
4320 INPUT"DISTANCE FROM PRODUCTION WELL (FT)=";R
4330 RA=M^2*BETA/R^2
4340 PRINT:PRINT"RATIO PV/PH=" USING"##.####^^^^";RA
4350 GOTO 5280
4360 PRINT:INPUT"AQUIFER TRANSMISSIVITY (GPD/FT)=";TRANS
4370 INPUT"DEVIATION OF TYPE CURVE TRACES DUE TO IMAGE WELL (FT)=";SD
4380 INPUT"DISCHARGE RATE (GPM)=";Q
4390 WUI=TRANS*SD/(114.6*Q)
4400 PRINT:PRINT"IMAGE WELL FUNCTION (DIM)=" USING"##.####^^^^";WUI
4410 GOTO 5280
4420 PRINT:INPUT"AQUIFER TRANSMISSIVITY=";TRANS
4430 INPUT"UI (DIM)=";UI
4440 INPUT"TIME (MIN)=";TIME
4450 INPUT"AQUIFER STORATIVITY (DIM)=";STOR
4460 RI=(TRANS*UI*TIME/(2693*STOR))^.5
4470 PRINT:PRINT"DISTANCE FROM OBSERVATION WELL TO IMAGE WELL (FT)=" USING"##.##
##^^^^";RI
4480 GOTO 5280
4490 PRINT:INPUT"AQUIFER TRANSMISSIVITY (GPD/FT)=";TRANS
```

```
4500 INPUT"DISTANCE FROM PRODUCTION WELL (FT)=";R
4510 INPUT"AQUIFER STORATIVITY (DIM)=";STOR
4520 TS=135000!*R^2*STOR/TRANS
4530 PRINT
4540 PRINT"TIME THAT MUST ELAPSE BEFORE U(=.02 (MIN)=" USING"######.##";TS
4550 GOTO 5280
4560 PRINT:INPUT"DISCHARGE RATE (GPM)=";Q
4570 INPUT"DRAWDOWN DIFFERENCE PER LOG CYCLE (FT)=";SL
4580 TRANS=264*Q/SL
4590 PRINT:PRINT"AQUIFER TRANSMISSIVITY (GPD/FT)=" USING"##.####^^^^";TRANS
4600 GOTO 5280
4610 PRINT:INPUT"AQUIFER TRANSMISSIVITY (GPD/FT)=";TRANS
4620 INPUT"ZERO-DRAWDOWN INTERCEPT (MIN)=";TI
4630 INPUT"DISTANCE FROM PRODUCTION WELL (FT)=";R
4640 STOR=TRANS*TI/(4790*R^2)
4650 PRINT:PRINT"AQUIFER STORATIVITY (DIM)=" USING"##.####^^^^";STOR
4660 GOTO 5280
4670 PRINT:INPUT"DISCHARGE RATE (GPM)=";Q
4680 INPUT"DRAWDOWN DIFFERENCE PER LOG CYCLE (FT)=";SL
4690 TRANS=528*Q/SL
4700 PRINT:PRINT"AQUIFER TRANSMISSIVITY (GPD/FT)=" USING"##.####^^^^";TRANS
4710 GOTO 5280
4720 PRINT:INPUT"AQUIFER TRANSMISSIVITY (GPD/FT)=";TRANS
4730 INPUT"TIME AFTER PUMPING STARTED (MIN)=";TIME
4740 INPUT"ZERO-DRAWDOWN INTERCEPT (FT)=";RO
4750 STOR=TRANS*TIME/(4790*RO^2)
4760 PRINT:PRINT"AQUIFER STORATIVITY (DIM)=" USING"##.####^^^^";STOR
4770 GOTO 5280
4780 PRINT:INPUT"DISTANCE FROM PRODUCTION WELL (FT)=";R
4790 INPUT"AQUIFER TRANSMISSIVITY (GPD/FT)=";TRANS
4800 INPUT"DRAWDOWN IN OBSERVATION WELL (FT)=";DS
4810 INPUT"DISCHARGE RATE (GPM)=";Q
4820 DI=2*((R^2*(10^(TRANS*DS/(528*Q)))^2-R^2)/4)^.5
4830 PRINT:PRINT"DISTANCE FROM PRODUCTION WELL"
4840 PRINT"TO RECHARGING IMAGE WELL (FT)=" USING"##.####^^^^";DI
4850 GOTO 5280
4860 PRINT:INPUT"RATE OF STREAM DEPLETION (GPM)=";QR
4870 INPUT"INDUCED STREAMBED INFILTRATION AREA (SQ FT)=";AR
4880 IA=6.3E+07*QR/AR
4890 PRINT:PRINT"INDUCED STREAMBED"
4900 PRINT"INFILTRATION RATE (GPD/ACRE)=" USING"##.####^^^^";IA
```

```
4910 GOTO 5280
4920 PRINT:INPUT"INDUCED STREAMBED INFILTRATION RATE (GPD/ACRE)=";IA
4930 INPUT"AVERAGE HEAD LOSS BENEATH STREAMBED (FT)=";HR
4940 IH=IA/HR
4950 PRINT:PRINT"INDUCED STREAMBED INFILTRATION RATE"
4960 PRINT"PER FOOT OF HEAD LOSS (GPD/ACRE/FT)=" USING"##.####^^^^";IH
4970 GOTO 5280
4980 PRINT:PRINT"INDUCED STREAMBED INFILTRATION RATE"
4990 INPUT"PER FOOT OF HEAD LOSS AT TEMP. DURING TEST (GPD/ACRE/FT)=";IH
5000 PRINT"SURFACE WATER DYNAMIC VISCOSITY"
5010 INPUT"AT TEMPERATURE DURING PUMPING TEST (POISE.SECOND)=";VA
5020 PRINT"SURFACE WATER DYNAMIC VISCOSITY"
5030 INPUT"AT SELECTED TEMPERATURE (POISE.SECOND)=";VT
5040 IT=IH*(VA/VT)
5050 PRINT:PRINT"INDUCED STREAMBED INFILTRATION RATE PER FOOT OF HEAD LOSS"
5060 PRINT"AT A SELECTED TEMPERATURE (GPD/ACRE/FT)=" USING"##.####^^^^";IT
5070 GOTO 5280
5080 PRINT:INPUT"WELL LOSS COEFFICIENT (SEC^2/FT^5)=";C
5090 INPUT"DISCHARGE RATE (CFS)=";Q
5100 SW=C*Q^2
5110 PRINT:PRINT"DRAWDOWN COMPONENT DUE TO WELL LOSS (FT)=" USING"#####.##";SW
5120 GOTO 5280
5130 PRINT:INPUT"INCREMENT OF DRAWDOWN FOR STEP 2 (FT)=";S2
5140 INPUT"INCREMENT OF DISCHARGE FOR STEP 2 (CFS)=";Q2
5150 INPUT"INCREMENT OF DRAWDOWN FOR STEP 1 (FT)=";S1
5160 INPUT"INCREMENT OF DISCHARGE FOR STEP 1 (CFS)=";Q1
5170 C=(S2/Q2-S1/Q1)/(Q1+Q2)
5180 PRINT:PRINT"WELL LOSS COEFFICIENT"
5190 PRINT"FOR STEPS 1&2 (SEC^2/FT^5)=" USING"######.##";C
5200 GOTO 5280
5210 PRINT:INPUT"INCREMENT OF DRAWDOWN FOR STEP 3 (FT)=";S3
5220 INPUT"INCREMENT OF DISCHARGE FOR STEP 3 (CFS)=";Q3
5230 INPUT"INCREMENT OF DRAWDOWN FOR STEP 2 (FT)=";S2
5240 INPUT"INCREMENT OF DISCHARGE FOR STEP 2 (CFS)=";Q2
5250 C=(S3/Q3-S2/Q2)/(Q2+Q3)
5260 PRINT:PRINT"WELL LOSS COEFFICIENT"
5270 PRINT"FOR STEPS 2&3 (SEC^2/FT^5)=" USING"######.##";C
5280 PRINT:PRINT"Enter Y for another equation"
5290 INPUT"or N to end program";E$
5300 IF E$<>"Y" AND E$<>"y" AND E$<>"N" AND E$<>"n" THEN 5280
5310 IF E$="Y" OR E$="y" THEN 70
5320 END
```

CATALOG LIST OF PROGRAMS ON DISKETTE

```
10 CLS:CLEAR:KEY OFF
20 PRINT"Program: CATALOG"
30 PRINT"Author : W.C. Walton"
40 PRINT"Version: IBM/PC 2.1; Copyright 1987 Lewis Publishers, Inc."
50 PRINT"Purpose: LIST PROGRAMS ON DISKETTE"
60 PRINT"         BY FILENAME AND PURPOSE":PRINT
70 PRINT"PROGRAM      PROGRAM"
80 PRINT"FILENAME     PURPOSE"
90 PRINT
100 PRINT"PT1          SIMULATION OF 1 OR 2-LAYER AQUIFER SYSTEM,"
110 PRINT"             UNIFORM PROPERTIES, WELL STORAGE CAPACITY,"
120 PRINT"             DELAYED GRAVITY YIELD, LEAKAGE, DEWATERING,"
130 PRINT"             RADIAL FLOW TO PRODUCTION WELL, FINITE-"
140 PRINT"             DIFFERENCE APPROXIMATION FOR PUMPING TEST"
150 PRINT"             DESIGN"
160 PRINT
170 PRINT"PT2          CALCULATE W(U,R/M(PV/PH)^.5/M,L/M,D/M,LO/M,DO/M)"
180 PRINT"             AND PARTIAL PENETRATION IMPACTS FOR PUMPING TEST"
190 PRINT"             ANALYSIS"
200 PRINT
210 PRINT"Press any key to continue"
220 A$=INKEY$:IF A$="" THEN 220
230 PRINT
240 PRINT"PT3          INTERPRET WELL FUNCTION VALUES BETWEEN VARIABLE"
250 PRINT"             VALUES GIVEN IN A TABLE FOR PUMPING TEST ANALYSIS"
260 PRINT
270 PRINT"PT4          CALCULATE W(U) + W(U,R/M(PV/PH)^.5,L/M,D/M,LO/M,DO/M)"
280 PRINT"             FOR PUMPING TEST ANALYSIS"
290 PRINT
300 PRINT"PT5          CALCULATE W(U,R/B) + W(U,R/M(PV/PH)^.5,L/M,D/M,"
310 PRINT"             LO/M,DO/M) FOR PUMPING TEST ANALYSIS"
320 PRINT
330 PRINT"PT6          CALCULATE AQUIFER TRANSMISSIVITY AND STORATIVITY"
340 PRINT"             FROM SEMILOG TIME- OR DISTANCE-DRAWDOWN GRAPH"
350 PRINT"             FOR PUMPING TEST ANALYSIS"
360 PRINT
370 PRINT"PT7          DETERMINE STORATIVITY BY METHOD OF SUCCESSIVE"
380 PRINT"             APPROXIMATIONS FOR PUMPING TEST ANALYSIS"
390 PRINT
```

```
400 PRINT"PT8          CALCULATE STREAM DEPLETION RATE FOR PUMPING"
410 PRINT"             TEST ANALYSIS"
420 PRINT
430 PRINT"PT9          CALCULATE DRAWDOWN BENEATH A STREAMBED WITH"
440 PRINT"             INDUCED INFILTRATION FOR PUMPING TEST ANALYSIS"
450 PRINT
460 PRINT"Press any key to continue"
470 A$=INKEY$:IF A$="" THEN 470
480 PRINT
490 PRINT"PT10         CALCULATE WELL LOSS COEFFICIENT FOR PUMPING"
500 PRINT"             TEST ANALYSIS"
510 PRINT
520 PRINT"PT11         SOLVE PUMPING TEST DESIGN AND ANALYSIS EQUATIONS"
530 PRINT
540 PRINT"Enter Y for hard copy of catalog or"
550 INPUT"N to end program";AF$
560 PRINT
570 IF AF$="N" OR AF$="n" THEN 1010
580 IF AF$()"N" AND AF$()"n" AND AF$()"Y" AND AF$()"y" THEN 540
590 LPRINT
600 LPRINT"PROGRAM          PROGRAM"
610 LPRINT"FILENAME         PURPOSE"
620 LPRINT
630 LPRINT"PT1           SIMULATION OF 1 OR 2-LAYER AQUIFER SYSTEM,"
640 LPRINT"              UNIFORM PROPERTIES, WELL STORAGE CAPACITY,"
650 LPRINT"              DELAYED GRAVITY YIELD, LEAKAGE, DEWATERING,"
660 LPRINT"              RADIAL FLOW TO PRODUCTION WELL, FINITE-"
670 LPRINT"              DIFFERENCE APPROXIMATION FOR PUMPING TEST"
680 LPRINT"              DESIGN"
690 LPRINT
700 LPRINT"PT2           CALCULATE W(U,R/M(PV/PH)^.5/M,L/M,D/M,LO/M,DO/M)"
710 LPRINT"              AND PARTIAL PENETRATION IMPACTS FOR PUMPING TEST"
720 LPRINT"              ANALYSIS"
730 LPRINT
740 LPRINT"PT3           INTERPRET WELL FUNCTION VALUES BETWEEN VARIABLE"
750 LPRINT"              VALUES GIVEN IN A TABLE FOR PUMPING TEST ANALYSIS"
760 LPRINT
770 LPRINT"PT4           CALCULATE W(U) + W(U,R/M(PV/PH)^.5,L/M,D/M,LO/M,DO/M)"
780 LPRINT"              FOR PUMPING TEST ANALYSIS"
790 LPRINT
800 LPRINT"PT5           CALCULATE W(U,R/B) + W(U,R/M(PV/PH)^.5,L/M,D/M,"
```

```
810 LPRINT"              LO/M,DO/M) FOR PUMPING TEST ANALYSIS"
820 LPRINT
830 LPRINT"PT6           CALCULATE AQUIFER TRANSMISSIVITY AND STORATIVITY"
840 LPRINT"              FROM SEMILOG TIME- OR DISTANCE-DRAWDOWN GRAPH"
850 LPRINT"              FOR PUMPING TEST ANALYSIS"
860 LPRINT
870 LPRINT"PT7           DETERMINE STORATIVITY BY METHOD OF SUCCESSIVE"
880 LPRINT"              APPROXIMATIONS FOR PUMPING TEST ANALYSIS"
890 LPRINT
900 LPRINT"PT8           CALCULATE STREAM DEPLETION RATE FOR PUMPING"
910 LPRINT"              TEST ANALYSIS"
920 LPRINT
930 LPRINT"PT9           CALCULATE DRAWDOWN BENEATH A STREAMBED WITH"
940 LPRINT"              INDUCED INFILTRATION FOR PUMPING TEST ANALYSIS"
950 LPRINT
960 LPRINT"PT10          CALCULATE WELL LOSS COEFFICIENT FOR PUMPING"
970 LPRINT"              TEST ANALYSIS"
980 LPRINT
990 LPRINT"PT11          SOLVE PUMPING TEST DESIGN AND ANALYSIS EQUATIONS"
1000 LPRINT
1010 END
```

Appendix B
Diskette Instructions

Instructions for using the GWPT diskette, containing 11 programs and a catalog of programs, are given in this appendix. It is assumed that the reader is familiar with the operation of the computer and its peripherals and understands the Disk Operating System manual supplied with the computer. Familiarity with groundwater model literature (see Walton, 1985) and the IBM PC BASIC language also is assumed.

These instructions were written for an IBM Personal Computer or "compatible" running PC-DOS or MS-DOS Version 2.0 or later (PC-DOS is a trademark of International Business Machines, Inc., and MS-DOS is a trademark of Microsoft Corp.) with 256K RAM available memory, a monochrome monitor, and a PC-DOS or MS-DOS "compatible" printer. Instructions assume you have two diskette drives; if you have one diskette drive or a fixed disk, consult your Disk Operating System manual.

The GWPT diskette does not contain DOS commands because of copyright restrictions prohibiting the distribution of bootable diskettes. Insert your DOS system diskette in drive A, insert the GWPT diskette in drive B, and turn on the power switch. Respond to prompts by entering the date and time. In response to the system prompt A> request a directory of GWPT using the DOS command: dir b:. Your disk should display the following GWPT files (though not necessarily in this order): CATALOG, PT1, PT2, PT3, PT4, PT5, PT6, PT7, PT8, PT9, PT10, and PT11. If all of these files are not on the GWPT diskette, contact the publisher. The contents of the program CATA-

LOG, which lists programs by filename and purpose, is as follows:

Filename	Purpose
PT1	1- or 2-layer aquifer system, uniform properties, well storage capacity, delayed gravity yield, leakage, dewatering, radial flow to production well, finite-difference approximation for pumping test design
PT2	Calculate $W[u,r(P_V/P_H)\hat{\ }0.5/m,l/m,d/m,l_o/m,d_o/m]$ and partial penetration impacts for pumping test analysis
PT3	Interpret well function values between variable values given in a table for pumping test analysis
PT4	Calculate $W(u) + W[u,r(P_V/P_H)\hat{\ }0.5/m,l/m,d/m,l_o/m,d_o/m]$ for pumping test analysis
PT5	Calculate $W(u,r/B) + W[u,r(P_V/P_H)\hat{\ }0.5/m,l/m,d/m,l_o/m,d_o/m]$ for pumping test analysis
PT6	Calculate aquifer transmissivity and storativity from semilog time- or distance-drawdown graph for pumping test analysis
PT7	Determine storativity by method of successive approximations for pumping test analysis
PT8	Calculate stream depletion rate for pumping test analysis
PT9	Calculate drawdown beneath a streambed with induced infiltration for pumping test analysis
PT10	Calculate well loss coefficient for pumping test analysis
PT11	Solve pumping test design and analysis equations

With your DOS system diskette in drive A, remove the GWPT diskette from drive B and insert a blank diskette in drive B. In response to the system prompt A > , format the blank diskette using the DOS command: format b:/s. Remove your DOS system diskette from drive A and insert the GWPT diskette in drive A. In response to the system

Instructions for BASIC programs contained in "Groundwater Pumping Tests," by William C. Walton, Copyright 1987 Lewis Publishers, Inc.

(Note: These programs need a printer in order to run.)

If your computer is an IBM-*compatible,* go to the DOS directory (or insert your DOS disk in A: drive if you do not have a hard disk) and type "gwbasic" and return. Insert the Walton BASIC disk in drive B: or copy the BASIC programs to your hard drive (usually drive C:).

(If your computer is a *genuine* IBM, type "basica" instead of "gwbasic.")

Hit the F3 function key and type the name of the program you wish to run, followed by a return. You must tell the computer where the BASIC program is—that is, supply a path name. Examples:

B:PT1 *or* C:\WALTON\PT1

Then hit the F2 function key to run the program.

When you are through with the program, you may either type "new" and then load and run another BASIC program as above, or you may type "system" to get back to the DOS prompt.

prompt A> , copy GWPT programs to the formatted working diskette in drive B using the following DOS command: copy *.* b:. Remove the GWPT diskette from drive A and insert your DOS system diskette in drive A. In response to the system prompt A> , copy the BASIC file onto the formatted working diskette in drive B using the DOS command: copy basic.com b:. The formatted working diskette's directory is now complete. To use the working diskette in drive A, in response to the system prompt A> get into BASIC by entering the command: basic. In response to the basic prompt ok, enter the load command with the desired GWPT program filename such as: load "b:pt1",r, and the specified program will be loaded and then run. Turn on the printer or else an error will occur.

Execution times for most programs will be a few minutes or less. Program PT1 may require more than an hour to run. Execution times may be shortened (speed-up factor of 3 to 20) by compiling programs with available software such as the QuickBASIC Compiler Version 4.0 or later (trademark of Microsoft Corp.) or Turbo Basic (trademark of Borland International, Inc.), which allow numerical processing in the 8087 math coprocessor environment. The information file "readme" may be displayed on the screen by entering "type readme" in response to the system prompt A>.

The user is interactively prompted as to when and what data should be entered with specified units. The prompt is in the form of a sentence written on the display screen. Output information is displayed at the screen and printer. Pre- or post-processor programs are not required. Programs are not menu driven.

The author has taken due care in preparing the GWPT diskette including research and testing to insure its accuracy and effectiveness. Neither the author nor the publisher makes any warranty of any kind, expressed or implied, with regard to the performance of the diskette and associated source codes. No warranties, expressed or implied, are made by the author or publisher that the programs are free of error, or are consistent with any particular standard of merchantability, or that they meet the reader's requirements for any particular application. In no event shall the author or the publisher be liable for incidental or consequential damages in connection with or arising from the furnishing, performance, or use of the programs on the GWPT diskette. Programs will give meaningful results only for reasonable problems. Support for the GWPT diskette and associated source codes is this book and is not otherwise available via telephone or mail.

Although this book and its programs are copyrighted, the reader is authorized to make one machine-readable copy of each program for personal use. Distribution of the machine-readable programs (either as copied by the reader or supplied with this book) is not authorized. Before use, programs should be verified with respect to known problem solutions over the range of available data.

Appendix C
Representative Hydraulic Characteristics

Representative aquifer system hydraulic characteristics and induced streambed infiltration rates presented by Walton (1985, pp. 19–23, 58) are listed as six tables in this appendix to assist the reader in quantifying pretest conceptual models. Values are given for aquifer horizontal hydraulic conductivity, aquitard vertical hydraulic conductivity, aquifer ratio of vertical and horizontal hydraulic conductivities, and rock-specific yield and artesian storativity.

Table C.1. Representative Horizontal Hydraulic Conductivity Values

Rock	Horizontal Hydraulic Conductivity (gpd/sq ft)
Gravel	1×10^{3} - 3×10^{4}
Basalt	1×10^{-6} - 2×10^{4}
Limestone	2×10^{-2} - 2×10^{4}
Sand and Gravel	2×10^{2} - 5×10^{3}
Sand	1×10^{2} - 3×10^{3}
Sand, quick	50 - 8×10^{3}
Sand, dune	1×10^{2} - 3×10^{2}
Peat	4 - 1×10^{2}
Sandstone	1×10^{-1} - 50
Loess	2×10^{-3} - 20
Clay	2×10^{-4} - 2
Till	5×10^{-4} - 1
Shale	1×10^{-5} - 1×10^{-1}
Coal	1 - 1×10^{3}

Table C.2. Representative Aquitard Vertical Hydraulic Conductivity

Rock	Aquitard Vertical Hydraulic Conductivity (gpd/sq ft)
Sand, gravel, and clay	1×10^{-1} - 1×10^{0}
Clay, sand, and gravel	1×10^{-2} - 6×10^{-2}
Clay	5×10^{-4} - 1×10^{-2}
Shale	1×10^{-7} - 1×10^{-3}

Table C.3. Representative Stratification Ratios

Degree of Stratification	P_V/P_H Ratio
Low	1/2
Medium	1/10
High	1/100

Table C.4. Representative Specific Yield Values

Rock	Specific Yield (dimensionless)
Peat	0.30 - 0.50
Sand, dune	0.30 - 0.40
Sand, coarse	0.20 - 0.35
Sand, gravelly	0.20 - 0.35
Gravel, fine	0.20 - 0.35
Gravel, coarse	0.10 - 0.25
Gravel, medium	0.15 - 0.25
Loess	0.15 - 0.35
Sand, medium	0.15 - 0.30
Sand, fine	0.10 - 0.30
Igneous, weathered	0.20 - 0.30
Sandstone	0.10 - 0.40
Sand and gravel	0.15 - 0.30
Silt	0.01 - 0.30
Clay, sandy	0.03 - 0.20
Clay	0.01 - 0.20
Volcanic, tuff	0.02 - 0.35
Siltstone	0.01 - 0.35
Limestone	0.01 - 0.25
Till	0.05 - 0.20

Table C.5. Representative Artesian Storativity Values

Material	Artesian Storativity (dimensionless)
Clay	1×10^{-4}m[a]
Sand and gravel	1×10^{-5}m
Rock, fissured	1×10^{-6}m

[a]m = aquifer or aquitard thickness (ft)

Table C.6. Representative Induced Streambed Infiltration Rates

Location	Induced Streambed Infiltration Rate (gpd/acre/ft)	Temperature of Surface Water (°F)
Mad River-Springfield, OH	1.0×10^{6}	39
Sandy Creek-Canton, OH	7.2×10^{5}	82
Mississippi River-St. Louis, IL	3.1×10^{5}	54
White River-Anderson, IN	2.2×10^{5}	69
Miami River-Cincinnati, OH	1.7×10^{5}	35
Mississippi River-St. Louis, IL	9.1×10^{4}	33
White River-Anderson, IN	4.0×10^{4}	38
Mississippi River-St. Louis, IL	3.5×10^{4}	83

Appendix D
Unit Conversion Factors

A set of factors for converting from the metric system of units to the gal-day-ft system of units and vice versa, presented by Walton (1985, pp. 577–583), is provided in this appendix to assist the reader in responding to the microcomputer program database prompts and translating computation results. Length, time, volumetric rate, hydraulic conductivity, and transmissivity factors are covered.

Length

To Convert from	to	Multiply by	Inverse
feet	meters	3.048×10^{-1}	3.2808
feet	miles	1.894×10^{-4}	5.380×10^{3}
feet	inches	1.200×10^{1}	8.333×10^{-2}
inches	meters	2.540×10^{-2}	3.937×10^{1}
inches	millimeters	2.540×10^{1}	3.937×10^{-2}
centimeters	meters	1.000×10^{-2}	1.000×10^{2}
centimeters	millimeters	1.000×10^{1}	1.000×10^{-1}
meters	miles	6.214×10^{-4}	1.609×10^{3}

Time

To Convert from	to	Multiply by	Inverse
days	seconds	8.640×10^{4}	1.157×10^{-5}
days	minutes	1.440×10^{3}	6.944×10^{-4}
days	hours	2.400×10^{1}	4.167×10^{-2}
hours	seconds	3.600×10^{3}	2.778×10^{-4}
hours	minutes	6.000×10^{1}	1.667×10^{-2}
minutes	seconds	6.000×10^{1}	1.667×10^{-2}

Volumetric Rate

To Convert from	to	Multiply by	Inverse
cu ft/min	gals(US)/sec	1.247×10^{-1}	8.019
cu ft/min	liters/sec	4.720×10^{-1}	2.119
cu ft/sec	gals(US)/min	4.488×10^{2}	2.228×10^{-3}
cu ft/sec	cu meters/sec	2.832×10^{-2}	3.531×10^{1}
cu ft/sec	liters/min	1.699×10^{3}	5.886×10^{-4}
cu ft/sec	gals(Imp)/min	3.737×10^{2}	2.676×10^{-3}
cu ft/sec	gals(US)/day	6.463×10^{5}	1.547×10^{-6}
cu ft/day	gals(US)/day	7.481	1.337×10^{-1}
cu ft/day	gals(Imp)/day	6.229	1.605×10^{-1}
cu meters/sec	cu ft/day	8.640×10^{4}	1.157×10^{-5}
cu meters/sec	gals(US)/day	2.282×10^{7}	4.382×10^{-8}
cu meters/sec	gals(Imp)/day	1.901×10^{7}	5.260×10^{-8}
cu meters/sec	cu ft/day	3.051×10^{6}	3.278×10^{-7}
gals(US)/min	gals(Imp)/min	8.327×10^{-1}	1.201
liters/sec	gals(US)/day	2.282×10^{4}	4.382×10^{-5}
liters/sec	gals(Imp)/day	1.901×10^{4}	5.260×10^{-5}
cu meters/day	gals(US)/min	1.840×10^{-1}	5.450

Hydraulic Conductivity

To Convert from	to	Multiply by	Inverse
gpd(US)/sq ft	ft/sec	1.743×10^{-6}	5.737×10^{5}
gpd(US)/sq ft	ft/day	1.340×10^{-1}	7.463
gpd(US)/sq ft	meters/day	4.075×10^{-2}	2.454×10^{1}
gpd(US)/sq ft	meters/min	2.830×10^{-5}	3.534×10^{4}
gpd(US)/sq ft	meters/sec	4.716×10^{-7}	2.120×10^{6}
gpd(Imp)/sq ft	meters/day	4.893×10^{-2}	2.044×10^{1}
gpd(Imp)/sq ft	meters/min	3.398×10^{-5}	2.943×10^{4}
gpd(Imp)/sq ft	meters/sec	5.663×10^{-7}	1.766×10^{6}
ft/day	meters/sec	3.528×10^{-6}	2.835×10^{5}
centimeter/sec	gals(US)/sq ft	2.121×10^{4}	4.716×10^{-5}

Transmissivity

To Convert from	to	Multiply by	Inverse
gpd(US)/ft	sq meters/day	1.242×10^{-2}	8.052×10^{1}
gpd(US)/ft	sq meters/min	8.624×10^{-6}	1.160×10^{5}
gpd(US)/ft	sq meters/sec	1.437×10^{-7}	6.959×10^{6}
gpd(Imp)/ft	sq meters/day	1.492×10^{-2}	6.705×10^{1}
gpd(Imp)/ft	sq meters/min	1.036×10^{-5}	9.654×10^{4}
gpd(Imp)/ft	sq meters/sec	1.726×10^{-7}	5.794×10^{6}
sq ft/day	sq meters/sec	1.075×10^{-6}	9.303×10^{5}

Appendix E
Well Function and Viscosity Values

Values of five well functions and the s_c/s ratio for analytical aquifer system models, in terms of the practical range of u, and other variables, are listed as tables in this appendix to assist the reader in developing type curves. The following well functions are covered: W(u), W(u,S,Rho), W(u,r/B), W(u,Gamma), and W(u_A,u_B,Beta). Values of water viscosity, in terms of the practical range of surface water temperature, also are listed in a table to assist the reader in adjusting induced streambed infiltration rates for temperature changes.

Table E.1. Values of W(u) (after Ferris et al., 1962, pp. 96–97)

u	W(u)	u	W(u)
1.0×10^{-8}	17.8435	2.5×10^{-7}	14.6246
1.5×10^{-8}	17.4380	3.0×10^{-7}	14.4423
2.0×10^{-8}	17.1503	3.5×10^{-7}	14.2881
2.5×10^{-8}	16.9272	4.0×10^{-7}	14.1546
3.0×10^{-8}	16.7449	5.0×10^{-7}	13.9314
3.5×10^{-8}	16.5907	6.0×10^{-7}	13.7491
4.0×10^{-8}	16.4572	7.0×10^{-7}	13.5950
5.0×10^{-8}	16.2340	8.0×10^{-7}	13.4614
6.0×10^{-8}	16.0517	9.0×10^{-7}	13.3437
7.0×10^{-8}	15.8976	1.0×10^{-6}	13.2383
8.0×10^{-8}	15.7640	1.5×10^{-6}	12.8328
9.0×10^{-8}	15.6462	2.0×10^{-6}	12.5451
1.0×10^{-7}	15.5409	2.5×10^{-6}	12.3220
1.5×10^{-7}	15.1354	3.0×10^{-6}	12.1397
2.0×10^{-7}	14.8477	3.5×10^{-6}	11.9855

Table E.1. Continued

u	W(u)	u	W(u)
4.0×10^{-6}	11.8520	7.0×10^{-3}	4.3916
5.0×10^{-6}	11.6280	8.0×10^{-3}	4.2591
6.0×10^{-6}	11.4465	9.0×10^{-3}	4.1423
7.0×10^{-6}	11.2924	1.0×10^{-2}	4.0379
8.0×10^{-6}	11.1589	1.5×10^{-2}	3.6374
9.0×10^{-6}	11.0411	2.0×10^{-2}	3.3547
1.0×10^{-5}	10.9357	2.5×10^{-2}	3.1365
1.5×10^{-5}	10.5303	3.0×10^{-2}	2.9591
2.0×10^{-5}	10.2426	3.5×10^{-2}	2.8099
2.5×10^{-5}	10.0194	4.0×10^{-2}	2.6813
3.0×10^{-5}	9.8371	5.0×10^{-2}	2.4679
3.5×10^{-5}	9.6830	6.0×10^{-2}	2.2953
4.0×10^{-5}	9.5495	7.0×10^{-2}	2.1508
5.0×10^{-5}	9.3263	8.0×10^{-2}	2.0269
6.0×10^{-5}	9.1440	9.0×10^{-2}	1.9187
7.0×10^{-5}	8.9899	1.0×10^{-1}	1.8229
8.0×10^{-5}	8.8563	1.5×10^{-1}	1.4645
9.0×10^{-5}	8.7386	2.0×10^{-1}	1.2227
1.0×10^{-4}	8.6332	2.5×10^{-1}	1.0443
1.5×10^{-4}	8.2278	3.0×10^{-1}	0.9057
2.0×10^{-4}	7.9402	3.5×10^{-1}	0.7942
2.5×10^{-4}	7.7172	4.0×10^{-1}	0.7024
3.0×10^{-4}	7.5348	5.0×10^{-1}	0.5598
3.5×10^{-4}	7.3807	6.0×10^{-1}	0.4544
4.0×10^{-4}	7.2472	7.0×10^{-1}	0.3738
5.0×10^{-4}	7.0242	8.0×10^{-1}	0.3106
6.0×10^{-4}	6.8420	9.0×10^{-1}	0.2602
7.0×10^{-4}	6.6879	1.0×10^{0}	0.2194
8.0×10^{-4}	6.5545	1.5×10^{0}	0.1000
9.0×10^{-4}	6.4368	2.0×10^{0}	0.0489
1.0×10^{-3}	6.3315	2.5×10^{0}	0.0249
1.5×10^{-3}	5.9266	3.0×10^{0}	0.0131
2.0×10^{-3}	5.6394	3.5×10^{0}	0.0070
2.5×10^{-3}	5.4167	4.0×10^{0}	0.0038
3.0×10^{-3}	5.2349	5.0×10^{0}	0.0011
3.5×10^{-3}	5.0813	6.0×10^{0}	0.0004
4.0×10^{-3}	4.9482	7.0×10^{0}	0.0001
5.0×10^{-3}	4.7261	8.0×10^{0}	
6.0×10^{-3}	4.5448	9.0×10^{0}	

Table E.2. Values of W(u,S,Rho) (after Reed, 1980, pp. 41–43)

Rho = 1			S		
u	1.0×10^{-1}	1.0×10^{-2}	1.0×10^{-3}	1.0×10^{-4}	1.0×10^{-5}
2.0×10^{0}	4.88×10^{-2}	4.99×10^{-3}	5.00×10^{-4}	5.00×10^{-5}	5.00×10^{-6}
1.0×10^{0}	9.19×10^{-2}	9.91×10^{-3}	9.99×10^{-4}	1.00×10^{-4}	1.00×10^{-5}
5.0×10^{-1}	1.77×10^{-1}	1.97×10^{-2}	2.00×10^{-3}	2.00×10^{-4}	2.00×10^{-5}
2.0×10^{-1}	4.06×10^{-1}	4.89×10^{-2}	4.99×10^{-3}	5.00×10^{-4}	5.00×10^{-5}
1.0×10^{-1}	7.34×10^{-1}	9.67×10^{-2}	9.97×10^{-3}	1.00×10^{-3}	1.00×10^{-4}
5.0×10^{-2}	1.26×10^{0}	1.90×10^{-1}	1.99×10^{-2}	2.00×10^{-3}	2.00×10^{-4}
2.0×10^{-2}	2.30×10^{0}	4.53×10^{-1}	4.95×10^{-2}	5.00×10^{-3}	5.00×10^{-4}
1.0×10^{-2}	3.28×10^{0}	8.52×10^{-1}	9.83×10^{-2}	9.98×10^{-3}	1.00×10^{-3}
5.0×10^{-3}	4.26×10^{0}	1.54×10^{0}	1.95×10^{-1}	1.99×10^{-2}	2.00×10^{-3}
2.0×10^{-3}	5.42×10^{0}	3.04×10^{0}	4.73×10^{-1}	4.97×10^{-2}	5.00×10^{-3}
1.0×10^{-3}	6.21×10^{0}	4.55×10^{0}	9.07×10^{-1}	9.90×10^{-2}	9.99×10^{-3}
5.0×10^{-4}	6.96×10^{0}	6.03×10^{0}	1.69×10^{0}	1.97×10^{-1}	2.00×10^{-2}
2.0×10^{-4}	7.87×10^{0}	7.56×10^{0}	3.52×10^{0}	4.81×10^{-1}	4.98×10^{-2}
1.0×10^{-4}	8.57×10^{0}	8.44×10^{0}	5.53×10^{0}	9.34×10^{-1}	9.93×10^{-2}
5.0×10^{-5}	9.32×10^{0}	9.23×10^{0}	7.63×10^{0}	1.77×10^{0}	1.98×10^{-1}
2.0×10^{-5}	1.02×10^{1}	1.02×10^{1}	9.68×10^{0}	3.83×10^{0}	4.86×10^{-1}
1.0×10^{-5}		1.09×10^{1}	1.07×10^{1}	6.25×10^{0}	9.49×10^{-1}
5.0×10^{-6}		1.16×10^{1}	1.15×10^{1}	8.99×10^{0}	1.82×10^{0}
2.0×10^{-6}		1.25×10^{1}	1.25×10^{1}	1.17×10^{1}	4.03×10^{0}
1.0×10^{-6}		1.32×10^{1}	1.32×10^{1}	1.29×10^{1}	6.78×10^{0}
5.0×10^{-7}			1.39×10^{1}	1.38×10^{1}	1.01×10^{1}
2.0×10^{-7}			1.48×10^{1}	1.48×10^{1}	1.37×10^{1}
1.0×10^{-7}			1.55×10^{1}	1.55×10^{1}	1.51×10^{1}
5.0×10^{-8}				1.62×10^{1}	1.61×10^{1}
2.0×10^{-8}				1.71×10^{1}	1.71×10^{1}
1.0×10^{-8}				1.78×10^{1}	1.78×10^{1}
5.0×10^{-9}					1.85×10^{1}
2.0×10^{-9}					1.94×10^{1}
1.0×10^{-9}					2.02×10^{1}
Rho = 100					
2.0×10^{0}	4.48×10^{-2}	3.44×10^{-2}	8.38×10^{-3}	9.56×10^{-4}	9.77×10^{-5}
1.0×10^{0}	2.14×10^{-1}	1.91×10^{-1}	7.56×10^{-2}	1.01×10^{-2}	1.04×10^{-3}
5.0×10^{-1}	5.55×10^{-1}	5.31×10^{-1}	3.23×10^{-1}	5.62×10^{-2}	6.02×10^{-3}
2.0×10^{-1}		1.20×10^{0}	1.02×10^{0}	3.04×10^{-1}	3.61×10^{-2}
1.0×10^{-1}		1.81×10^{0}	1.70×10^{0}	7.92×10^{-1}	1.10×10^{-1}
5.0×10^{-2}		2.46×10^{0}	2.40×10^{0}	1.62×10^{0}	2.92×10^{-1}
2.0×10^{-2}		3.35×10^{0}	3.32×10^{0}	2.95×10^{0}	8.91×10^{-1}
1.0×10^{-2}			4.02×10^{0}	3.84×10^{0}	1.80×10^{0}
5.0×10^{-3}			4.72×10^{0}	4.63×10^{0}	3.14×10^{0}
2.0×10^{-3}			5.64×10^{0}	5.60×10^{0}	5.01×10^{0}
1.0×10^{-3}				6.31×10^{0}	6.06×10^{0}
5.0×10^{-4}				7.01×10^{0}	6.90×10^{0}
2.0×10^{-4}				7.94×10^{0}	7.89×10^{0}
1.0×10^{-4}					8.61×10^{0}
5.0×10^{-5}					9.31×10^{0}
2.0×10^{-5}					1.02×10^{1}

Table E.3. Values of W(u,r/B) (after Hantush, 1964, pp. 322–324)

u	r/B				
	1.0×10^{-3}	5.0×10^{-3}	1.0×10^{-2}	2.5×10^{-2}	5.0×10^{-2}
1.0×10^{-6}	13.0031	10.8283	9.4425	7.6111	6.2285
2.0×10^{-6}	12.4240	10.8174	9.4425	7.6111	6.2285
3.0×10^{-6}	12.0581	10.7849	9.4425	7.6111	6.2285
4.0×10^{-6}	11.7905	10.7374	9.4422	7.6111	6.2285
5.0×10^{-6}	11.5795	10.6822	9.4413	7.6111	6.2285
6.0×10^{-6}	11.4053	10.6240	9.4394	7.6111	6.2285
7.0×10^{-6}	11.2570	10.5652	9.4361	7.6111	6.2285
8.0×10^{-6}	11.1279	10.5072	9.4313	7.6111	6.2285
9.0×10^{-6}	11.0135	10.4508	9.4251	7.6111	6.2285
1.0×10^{-5}	10.9109	10.3963	9.4176	7.6111	6.2285
2.0×10^{-5}	10.2301	9.9530	9.2961	7.6111	6.2285
3.0×10^{-5}	9.8288	9.6392	9.1499	7.6101	6.2285
4.0×10^{-5}	9.5432	9.3992	9.0102	7.6069	6.2285
5.0×10^{-5}	9.3213	9.2052	8.8827	7.6000	6.2285
6.0×10^{-5}	9.1398	9.0426	8.7673	7.5894	6.2285
7.0×10^{-5}	8.9863	8.9027	8.6625	7.5754	6.2285
8.0×10^{-5}	8.8532	8.7798	8.5669	7.5589	6.2284
9.0×10^{-5}	8.7358	8.6703	8.4792	7.5402	6.2283
1.0×10^{-4}	8.6308	8.5717	8.3983	7.5199	6.2282
2.0×10^{-4}	7.9390	7.9092	7.8192	7.2898	6.2173
3.0×10^{-4}	7.5340	7.5141	7.4534	7.0759	6.1848
4.0×10^{-4}	7.2466	7.2317	7.1859	6.8929	6.1373
5.0×10^{-4}	7.0237	7.0118	6.9750	6.7357	6.0821
6.0×10^{-4}	6.8416	6.8316	6.8009	6.5988	6.0239
7.0×10^{-4}	6.6876	6.6790	6.6527	6.4777	5.9652
8.0×10^{-4}	6.5542	6.5467	6.5237	6.3695	5.9073
9.0×10^{-4}	6.4365	6.4299	6.4094	6.2716	5.8509
1.0×10^{-3}	6.3313	6.3253	6.3069	6.1823	5.7965
2.0×10^{-3}	5.6393	5.6363	5.6271	5.5638	5.3538
3.0×10^{-3}	5.2348	5.2329	5.2267	5.1845	5.0408
4.0×10^{-3}	4.9482	4.9467	4.9421	4.9105	4.8016
5.0×10^{-3}	4.7260	4.7249	4.7212	4.6960	4.6084
6.0×10^{-3}	4.5448	4.5438	4.5407	4.5197	4.4467
7.0×10^{-3}	4.3916	4.3908	4.3882	4.3702	4.3077
8.0×10^{-3}	4.2590	4.2583	4.2561	4.2404	4.1857
9.0×10^{-3}	4.1423	4.1416	4.1396	4.1258	4.0772
1.0×10^{-2}	4.0379	4.0373	4.0356	4.0231	3.9795
2.0×10^{-2}	3.3547	3.3544	3.3536	3.3476	3.3264
3.0×10^{-2}	2.9591	2.9589	2.9584	2.9545	2.9409
4.0×10^{-2}	2.6812	2.6811	2.6807	2.6779	2.6680
5.0×10^{-2}	2.4679	2.4678	2.4675	2.4653	2.4576
6.0×10^{-2}	2.2953	2.2952	2.2950	2.2932	2.2870
7.0×10^{-2}	2.1508	2.1508	2.1506	2.1491	2.1439
8.0×10^{-2}	2.0269	2.0269	2.0267	2.0255	2.0210

Table E.3. Continued

u	r/B 1.0×10^{-3}	5.0×10^{-3}	1.0×10^{-2}	2.5×10^{-2}	5.0×10^{-2}
9.0×10^{-2}	1.9187	1.9187	1.9185	1.9174	1.9136
1.0×10^{-1}	1.8229	1.8229	1.8227	1.8218	1.8184
2.0×10^{-1}	1.2226	1.2226	1.2226	1.2222	1.2209
3.0×10^{-1}	0.9057	0.9057	0.9056	0.9054	0.9047
4.0×10^{-1}	0.7024	0.7024	0.7024	0.7022	0.7016
5.0×10^{-1}	0.5598	0.5598	0.5598	0.5597	0.5594
6.0×10^{-1}	0.4544	0.4544	0.4544	0.4543	0.4541
7.0×10^{-1}	0.3738	0.3738	0.3738	0.3737	0.3735
8.0×10^{-1}	0.3106	0.3106	0.3106	0.3106	0.3104
9.0×10^{-1}	0.2602	0.2602	0.2602	0.2602	0.2601
1.0×10^{0}	0.2194	0.2194	0.2194	0.2194	0.2193
2.0×10^{0}	0.0489	0.0489	0.0489	0.0489	0.0489
3.0×10^{0}	0.0130	0.0130	0.0130	0.0130	0.0130
4.0×10^{0}	0.0038	0.0038	0.0038	0.0038	0.0038
5.0×10^{0}	0.0011	0.0011	0.0011	0.0011	0.0011
6.0×10^{0}	0.0004	0.0004	0.0004	0.0004	0.0004
7.0×10^{0}	0.0001	0.0001	0.0001	0.0001	0.0001

u	r/B 7.5×10^{-2}	1.5×10^{-1}	3.0×10^{-1}	5.0×10^{-1}	7.0×10^{-1}
1.0×10^{-4}	5.4228	4.0601	2.7449	1.8488	1.3210
2.0×10^{-4}	5.4227	4.0601	2.7449	1.8488	1.3210
3.0×10^{-4}	5.4212	4.0601	2.7449	1.8488	1.3210
4.0×10^{-4}	5.4160	4.0601	2.7449	1.8488	1.3210
5.0×10^{-4}	5.4062	4.0601	2.7449	1.8488	1.3210
6.0×10^{-4}	5.3921	4.0601	2.7449	1.8488	1.3210
7.0×10^{-4}	5.3745	4.0600	2.7449	1.8488	1.3210
8.0×10^{-4}	5.3542	4.0599	2.7449	1.8488	1.3210
9.0×10^{-4}	5.3317	4.0598	2.7449	1.8488	1.3210
1.0×10^{-3}	5.3078	4.0595	2.7449	1.8488	1.3210
2.0×10^{-3}	5.0517	4.0435	2.7449	1.8488	1.3210
3.0×10^{-3}	4.8243	4.0092	2.7448	1.8488	1.3210
4.0×10^{-3}	4.6335	3.9551	2.7444	1.8488	1.3210
5.0×10^{-3}	4.4713	3.8821	2.7428	1.8488	1.3210
6.0×10^{-3}	4.3311	3.8384	2.7398	1.8488	1.3210
7.0×10^{-3}	4.2078	3.7529	2.7350	1.8488	1.3210
8.0×10^{-3}	4.0980	3.6903	2.7284	1.8488	1.3210
9.0×10^{-3}	3.9991	3.6302	2.7202	1.8487	1.3210
1.0×10^{-2}	3.9091	3.5725	2.7104	1.8486	1.3210
2.0×10^{-2}	3.2917	3.1158	2.5688	1.8379	1.3207
3.0×10^{-2}	2.9183	2.8017	2.4110	1.8062	1.3177
4.0×10^{-2}	2.6515	2.5655	2.2661	1.7603	1.3094
5.0×10^{-2}	2.4448	2.3776	2.1371	1.7075	1.2955
6.0×10^{-2}	2.2766	2.2218	2.0227	1.6524	1.2770
7.0×10^{-2}	2.1352	2.0894	1.9206	1.5973	1.2551

Table E.3. Continued

u	r/B 7.5×10^{-2}	1.5×10^{-1}	3.0×10^{-1}	5.0×10^{-1}	7.0×10^{-1}
8.0×10^{-2}	2.0136	1.9745	1.8290	1.5436	1.2310
9.0×10^{-2}	1.9072	1.8732	1.7460	1.4918	1.2054
1.0×10^{-1}	1.8128	1.7829	1.6704	1.4422	1.1791
2.0×10^{-1}	1.2186	1.2066	1.1602	1.0592	0.9629
3.0×10^{-1}	0.9035	0.8969	0.8713	0.8142	0.7362
4.0×10^{-1}	0.7010	0.6969	0.6809	0.6446	0.5943
5.0×10^{-1}	0.5588	0.5561	0.5453	0.5206	0.4860
6.0×10^{-1}	0.4537	0.4518	0.4441	0.4266	0.4018
7.0×10^{-1}	0.3733	0.3719	0.3663	0.3534	0.3351
8.0×10^{-1}	0.3102	0.3092	0.3050	0.2953	0.2815
9.0×10^{-1}	0.2599	0.2591	0.2559	0.2485	0.2378
1.0×10^{0}	0.2191	0.2186	0.2161	0.2103	0.2020
2.0×10^{0}	0.0489	0.0488	0.0485	0.0477	0.0467
3.0×10^{0}	0.0130	0.0130	0.0130	0.0128	0.0126
4.0×10^{0}	0.0038	0.0038	0.0038	0.0037	0.0037
5.0×10^{0}	0.0011	0.0011	0.0011	0.0011	0.0011
6.0×10^{0}	0.0004	0.0004	0.0004	0.0004	0.0004
7.0×10^{0}	0.0001	0.0001	0.0001	0.0001	0.0001
u	**r/B** 8.5×10^{-1}	1.0×10^{0}	1.5×10^{0}	2.0×10^{0}	2.5×10^{0}
1.0×10^{-2}	1.0485	0.8420	0.4276	0.2278	0.1247
2.0×10^{-2}	1.0484	0.8420	0.4276	0.2278	0.1247
3.0×10^{-2}	1.0481	0.8420	0.4276	0.2278	0.1247
4.0×10^{-2}	1.0465	0.8418	0.4276	0.2278	0.1247
5.0×10^{-2}	1.0426	0.8409	0.4276	0.2278	0.1247
6.0×10^{-2}	1.0362	0.8391	0.4276	0.2278	0.1247
7.0×10^{-2}	1.0272	0.8360	0.4276	0.2278	0.1247
8.0×10^{-2}	1.0161	0.8316	0.4275	0.2278	0.1247
9.0×10^{-2}	1.0032	0.8259	0.4274	0.2278	0.1247
1.0×10^{-1}	0.9890	0.8190	0.4271	0.2278	0.1247
2.0×10^{-1}	0.8216	0.7148	0.4135	0.2268	0.1247
3.0×10^{-1}	0.6706	0.6010	0.3812	0.2211	0.1240
4.0×10^{-1}	0.5501	0.5024	0.3411	0.2096	0.1217
5.0×10^{-1}	0.4550	0.4210	0.3007	0.1944	0.1174
6.0×10^{-1}	0.3793	0.3543	0.2630	0.1774	0.1112
7.0×10^{-1}	0.3183	0.2996	0.2292	0.1602	0.1040
8.0×10^{-1}	0.2687	0.2543	0.1994	0.1436	0.0961
9.0×10^{-1}	0.2280	0.2168	0.1734	0.1281	0.0881
1.0×10^{0}	0.1943	0.1855	0.1509	0.1139	0.0803
2.0×10^{0}	0.0456	0.0444	0.0394	0.0335	0.0271
3.0×10^{0}	0.0124	0.0122	0.0112	0.0100	0.0086
4.0×10^{0}	0.0036	0.0036	0.0034	0.0031	0.0027
5.0×10^{0}	0.0011	0.0011	0.0010	0.0010	0.0009
6.0×10^{0}	0.0004	0.0004	0.0003	0.0003	0.0003
7.0×10^{0}	0.0001	0.0001	0.0001	0.0001	0.0001

Table E.4. Values of W(u,Gamma) (after Hantush, 1964, p. 313)

			Gamma		
u	1.0×10^{-2}	5.0×10^{-2}	1.0×10^{-1}	2.0×10^{-1}	5.0×10^{-1}
1.0×10^{-6}	9.9259	8.3395	7.6497	6.9590	6.0463
2.0×10^{-6}	9.5677	7.9908	7.3024	6.6126	5.7012
3.0×10^{-6}	9.3561	7.7864	7.0991	6.4100	5.4996
4.0×10^{-6}	9.2047	7.6412	6.9547	6.2663	5.3567
5.0×10^{-6}	9.0866	7.5284	6.8427	6.1548	5.2459
6.0×10^{-6}	8.9894	7.4362	6.7512	6.0637	5.1555
7.0×10^{-6}	8.9069	7.3581	6.6737	5.9867	5.0790
8.0×10^{-6}	8.8350	7.2904	6.6066	5.9200	5.0129
9.0×10^{-6}	8.7714	7.2306	6.5474	5.8611	4.9545
1.0×10^{-5}	8.7142	7.1771	6.4944	5.8085	4.9024
2.0×10^{-5}	8.3315	6.8238	6.1453	5.4623	4.5598
3.0×10^{-5}	8.1013	6.6159	5.9406	5.2597	4.3600
4.0×10^{-5}	7.9346	6.4677	5.7951	5.1160	4.2185
5.0×10^{-5}	7.8031	6.3523	5.6821	5.0045	4.1090
6.0×10^{-5}	7.6941	6.2576	5.5896	4.9134	4.0196
7.0×10^{-5}	7.6007	6.1773	5.5113	4.8364	3.9442
8.0×10^{-5}	7.5190	6.1076	5.4434	4.7697	3.8789
9.0×10^{-5}	7.4461	6.0459	5.3834	4.7108	3.8214
1.0×10^{-4}	7.3803	5.9906	5.3297	4.6581	3.7700
2.0×10^{-4}	6.9321	5.6226	4.9747	4.3115	3.4334
3.0×10^{-4}	6.6563	5.4035	4.7655	4.1086	3.2379
4.0×10^{-4}	6.4541	5.2459	4.6161	3.9645	3.0999
5.0×10^{-4}	6.2934	5.1223	4.4996	3.8527	2.9933
6.0×10^{-4}	6.1596	5.0203	4.4040	3.7612	2.9065
7.0×10^{-4}	6.0447	4.9333	4.3228	3.6838	2.8334
8.0×10^{-4}	5.9439	4.8573	4.2523	3.6167	2.7702
9.0×10^{-4}	5.8539	4.7898	4.1898	3.5575	2.7146
1.0×10^{-3}	5.7727	4.7290	4.1337	3.5045	2.6650
2.0×10^{-3}	5.2203	4.3184	3.7598	3.1549	2.3419
3.0×10^{-3}	4.8837	4.0683	3.5363	2.9494	2.1559
4.0×10^{-3}	4.6396	3.8859	3.3750	2.8030	2.0253
5.0×10^{-3}	4.4474	3.7415	3.2483	2.6891	1.9250
6.0×10^{-3}	4.2888	3.6214	3.1436	2.5957	1.8437
7.0×10^{-3}	4.1536	3.5185	3.0542	2.5165	1.7754
8.0×10^{-3}	4.0357	3.4282	2.9762	2.4478	1.7166
9.0×10^{-3}	3.9313	3.3478	2.9068	2.3870	1.6651
1.0×10^{-2}	3.8374	3.2752	2.8443	2.3325	1.6193
2.0×10^{-2}	3.2133	2.7829	2.4227	1.9714	1.3239
3.0×10^{-2}	2.8452	2.4844	2.1680	1.7579	1.1570
4.0×10^{-2}	2.5842	2.2691	1.9841	1.6056	1.0416
5.0×10^{-2}	2.3826	2.1007	1.8401	1.4872	0.9540

Table E.4. Continued

u	Gamma 1.0×10^{-2}	5.0×10^{-2}	1.0×10^{-1}	2.0×10^{-1}	5.0×10^{-1}
6.0×10^{-2}	2.2188	1.9626	1.7217	1.3905	0.8838
7.0×10^{-2}	2.0812	1.8458	1.6213	1.3088	0.8255
8.0×10^{-2}	1.9630	1.7448	1.5343	1.2381	0.7758
9.0×10^{-2}	1.8595	1.6559	1.4577	1.1760	0.7327
1.0×10^{-1}	1.7677	1.5768	1.3893	1.1207	0.6947
2.0×10^{-1}	1.1895	1.0714	0.9497	0.7665	0.4603
3.0×10^{-1}	0.8825	0.7986	0.7103	0.5739	0.3390
4.0×10^{-1}	0.6850	0.6218	0.5543	0.4482	0.2619
5.0×10^{-1}	0.5463	0.4969	0.4436	0.3591	0.2083
6.0×10^{-1}	0.4437	0.4041	0.3613	0.2927	0.1688
7.0×10^{-1}	0.3651	0.3330	0.2980	0.2415	0.1386
8.0×10^{-1}	0.3035	0.2770	0.2481	0.2012	0.1151
9.0×10^{-1}	0.2543	0.2323	0.2082	0.1690	0.0010
1.0×10^{0}	0.2144	0.1961	0.1758	0.1427	0.0008
2.0×10^{0}	0.0005	0.0004	0.0004	0.0003	0.0002
3.0×10^{0}	0.0001	0.0001	0.0001	0.0001	

u	Gamma 1.0×10^{0}	2.0×10^{0}	5.0×10^{0}	1.0×10^{1}	2.0×10^{1}
1.0×10^{-6}	5.3575	4.6721	3.7756	3.1110	2.4671
2.0×10^{-6}	5.0141	4.3312	3.4412	2.7857	2.1568
3.0×10^{-6}	4.8136	4.1327	3.2474	2.5984	1.9801
4.0×10^{-6}	4.6716	3.9922	3.1109	2.4671	1.8571
5.0×10^{-6}	4.5617	3.8836	3.0055	2.3661	1.7633
6.0×10^{-6}	4.4719	3.7951	2.9199	2.2844	1.6877
7.0×10^{-6}	4.3962	3.7204	2.8478	2.2158	1.6246
8.0×10^{-6}	4.3306	3.6558	2.7856	2.1568	1.5706
9.0×10^{-6}	4.2728	3.5989	2.7309	2.1050	1.5234
1.0×10^{-5}	4.2212	3.5481	2.6822	2.0590	1.4816
2.0×10^{-5}	3.8827	3.2162	2.3660	1.7632	1.2170
3.0×10^{-5}	3.6858	3.0241	2.1850	1.5965	1.0716
4.0×10^{-5}	3.5468	2.8889	2.0588	1.4815	0.9730
5.0×10^{-5}	3.4394	2.7848	1.9622	1.3943	0.8994
6.0×10^{-5}	3.3519	2.7002	1.8841	1.3244	0.8412
7.0×10^{-5}	3.2781	2.6290	1.8189	1.2664	0.7934
8.0×10^{-5}	3.2143	2.5677	1.7629	1.2169	0.7530
9.0×10^{-5}	3.1583	2.5138	1.7139	1.1739	0.7182
1.0×10^{-4}	3.1082	2.4658	1.6704	1.1359	0.6878
2.0×10^{-4}	2.7819	2.1549	1.3937	0.8992	0.5044
3.0×10^{-4}	2.5937	1.9778	1.2401	0.7721	0.4111
4.0×10^{-4}	2.4617	1.8545	1.1352	0.6875	0.3514
5.0×10^{-4}	2.3601	1.7604	1.0564	0.6252	0.3089

Table E.4. Continued

u	Gamma 1.0×10^{0}	2.0×10^{0}	5.0×10^{0}	1.0×10^{1}	2.0×10^{1}
6.0×10^{-4}	2.2778	1.6846	0.9937	0.5765	0.2766
7.0×10^{-4}	2.2087	1.6212	0.9420	0.5370	0.2510
8.0×10^{-4}	2.1492	1.5670	0.8982	0.5040	0.2300
9.0×10^{-4}	2.0971	1.5196	0.8603	0.4758	0.2125
1.0×10^{-3}	2.0506	1.4776	0.8271	0.4513	0.1976
2.0×10^{-3}	1.7516	1.2116	0.6238	0.3084	0.1164
3.0×10^{-3}	1.5825	1.0652	0.5182	0.2394	0.0008
4.0×10^{-3}	1.4656	0.9658	0.4496	0.1970	0.0006
5.0×10^{-3}	1.3767	0.8915	0.4001	0.1677	0.0005
6.0×10^{-3}	1.3054	0.8327	0.3620	0.1460	0.0004
7.0×10^{-3}	1.2460	0.7843	0.3315	0.1292	0.0003
8.0×10^{-3}	1.1953	0.7435	0.3064	0.1158	0.0003
9.0×10^{-3}	1.1512	0.7083	0.2852	0.1047	0.0003
1.0×10^{-2}	1.1122	0.6775	0.2670	0.0010	0.0002
2.0×10^{-2}	0.8677	0.4910	0.1653	0.0005	0.0001
3.0×10^{-2}	0.7353	0.3965	0.1197	0.0004	
4.0×10^{-2}	0.6467	0.3357	0.0009	0.0003	
5.0×10^{-2}	0.5812	0.2923	0.0008	0.0002	
6.0×10^{-2}	0.5298	0.2593	0.0006	0.0001	
7.0×10^{-2}	0.4880	0.2332	0.0005	0.0001	
8.0×10^{-2}	0.4530	0.2119	0.0005	0.0001	
9.0×10^{-2}	0.4230	0.1941	0.0004	0.0001	
1.0×10^{-1}	0.3970	0.1789	0.0004	0.0001	
2.0×10^{-1}	0.2452	0.0010	0.0001		
3.0×10^{-1}	0.1729	0.0006	0.0001		
4.0×10^{-1}	0.1296	0.0004	0.0001		
5.0×10^{-1}	0.1006	0.0003			
6.0×10^{-1}	0.0008	0.0002			
7.0×10^{-1}	0.0006	0.0002			
8.0×10^{-1}	0.0005	0.0002			
9.0×10^{-1}	0.0004	0.0001			
1.0×10^{0}	0.0004	0.0001			
2.0×10^{0}	0.0001	0.0001			
3.0×10^{0}	0.0001				

Table E.5. Values of s_c/s (after Witherspoon and Neuman, 1972, p. 267)

	u				
u_c	2.5×10^{-1}	2.5×10^{-2}	2.5×10^{-3}	2.5×10^{-4}	2.5×10^{-5}
2.0×10^{-4}	0.9710	0.9780	0.9800	0.9810	0.9820
4.0×10^{-4}	0.9580	0.9680	0.9720	0.9730	0.9740
6.0×10^{-4}	0.9490	0.9610	0.9660	0.9680	0.9690
8.0×10^{-4}	0.9420	0.9550	0.9600	0.9630	0.9640
2.0×10^{-3}	0.9090	0.9300	0.9370	0.9410	0.9430
4.0×10^{-3}	0.8730	0.9010	0.9120	0.9170	0.9200
6.0×10^{-3}	0.8460	0.8800	0.8930	0.8990	0.9020
8.0×10^{-3}	0.8240	0.8620	0.8770	0.8830	0.8870
2.0×10^{-2}	0.7320	0.7870	0.8080	0.8180	0.8230
4.0×10^{-2}	0.6370	0.7070	0.7340	0.7470	0.7540
6.0×10^{-2}	0.5710	0.6490	0.6800	0.6940	0.7020
8.0×10^{-2}	0.5200	0.6030	0.6360	0.6510	0.6600
2.0×10^{-1}	0.3360	0.4260	0.4640	0.4830	0.4930
4.0×10^{-1}	0.1950	0.2750	0.3110	0.3280	0.3380
6.0×10^{-1}	0.1230	0.1890	0.2200	0.2360	0.2440
8.0×10^{-1}	0.0818	0.1350	0.1610	0.1740	0.1810
2.0×10^{0}	0.0103	0.0233	0.0311	0.0352	0.0376
4.0×10^{0}	0.0001	0.0018	0.0028	0.0033	0.0036
6.0×10^{0}		0.0002	0.0003	0.0004	0.0004

Table E.6. Values of $W(u_A,u_B,Beta)$[a] (after Neuman, 1975, pp. 332–333)

$1/u_A$	Beta 1.0×10^{-3}	Beta 1.0×10^{-2}	Beta 6.0×10^{-2}	Beta 2.0×10^{-1}
4.0×10^{-1}	2.48×10^{-2}	2.41×10^{-2}	2.30×10^{-2}	2.14×10^{-2}
8.0×10^{-1}	1.45×10^{-1}	1.40×10^{-1}	1.31×10^{-1}	1.19×10^{-1}
1.4×10^{0}	3.58×10^{-1}	3.45×10^{-1}	3.18×10^{-1}	2.79×10^{-1}
2.4×10^{0}	6.62×10^{-1}	6.33×10^{-1}	5.70×10^{-1}	4.83×10^{-1}
4.0×10^{0}	1.02×10^{0}	9.63×10^{-1}	8.49×10^{-1}	6.88×10^{-1}
8.0×10^{0}	1.57×10^{0}	1.46×10^{0}	1.23×10^{0}	9.18×10^{-1}
1.4×10^{1}	2.05×10^{0}	1.88×10^{0}	1.51×10^{0}	1.03×10^{0}
2.4×10^{1}	2.52×10^{0}	2.27×10^{0}	1.73×10^{0}	1.07×10^{0}
4.0×10^{1}	2.97×10^{0}	2.61×10^{0}	1.85×10^{0}	1.08×10^{0}
8.0×10^{1}	3.56×10^{0}	3.00×10^{0}	1.92×10^{0}	1.08×10^{0}
1.4×10^{2}	4.01×10^{0}	3.23×10^{0}	1.93×10^{0}	1.08×10^{0}
2.4×10^{2}	4.42×10^{0}	3.37×10^{0}	1.94×10^{0}	1.08×10^{0}
4.0×10^{2}	4.77×10^{0}	3.43×10^{0}	1.94×10^{0}	1.08×10^{0}
8.0×10^{2}	5.16×10^{0}	3.45×10^{0}	1.94×10^{0}	1.08×10^{0}
1.4×10^{3}	5.40×10^{0}	3.46×10^{0}	1.94×10^{0}	1.08×10^{0}
2.4×10^{3}	5.54×10^{0}	3.46×10^{0}	1.94×10^{0}	1.08×10^{0}
4.0×10^{3}	5.59×10^{0}	3.46×10^{0}	1.94×10^{0}	1.08×10^{0}
8.0×10^{3}	5.62×10^{0}	3.46×10^{0}	1.94×10^{0}	1.08×10^{0}
1.4×10^{4}	5.62×10^{0}	3.46×10^{0}	1.94×10^{0}	1.08×10^{0}

$1/u_A$	Beta 6.0×10^{-1}	Beta 1.0×10^{0}	Beta 2.0×10^{0}	Beta 4.0×10^{0}
4.0×10^{-1}	1.88×10^{-2}	1.70×10^{-2}	1.38×10^{-2}	9.33×10^{-3}
8.0×10^{-1}	9.88×10^{-2}	8.49×10^{-2}	6.03×10^{-2}	3.17×10^{-2}
1.4×10^{0}	2.17×10^{-1}	1.75×10^{-1}	1.07×10^{-1}	4.45×10^{-2}
2.4×10^{0}	3.43×10^{-1}	2.56×10^{-1}	1.33×10^{-1}	4.76×10^{-2}
4.0×10^{0}	4.38×10^{-1}	3.00×10^{-1}	1.40×10^{-1}	4.78×10^{-2}
8.0×10^{0}	4.97×10^{-1}	3.17×10^{-1}	1.41×10^{-1}	4.78×10^{-2}
1.4×10^{1}	5.07×10^{-1}	3.17×10^{-1}	1.41×10^{-1}	4.78×10^{-2}
2.4×10^{1}	5.07×10^{-1}	3.17×10^{-1}	1.41×10^{-1}	4.78×10^{-2}

$1/u_B$	Beta 1.0×10^{-3}	Beta 1.0×10^{-2}	Beta 6.0×10^{-2}	Beta 2.0×10^{-1}
4.0×10^{-2}	5.62×10^{0}	3.46×10^{0}	1.94×10^{0}	1.09×10^{0}
8.0×10^{-2}	5.62×10^{0}	3.46×10^{0}	1.94×10^{0}	1.09×10^{0}
1.4×10^{-1}	5.62×10^{0}	3.46×10^{0}	1.94×10^{0}	1.10×10^{0}
2.4×10^{-1}	5.62×10^{0}	3.46×10^{0}	1.95×10^{0}	1.11×10^{0}
4.0×10^{-1}	5.62×10^{0}	3.46×10^{0}	1.96×10^{0}	1.13×10^{0}
8.0×10^{-1}	5.62×10^{0}	3.46×10^{0}	1.98×10^{0}	1.18×10^{0}
1.4×10^{0}	5.63×10^{0}	3.47×10^{0}	2.01×10^{0}	1.24×10^{0}
2.4×10^{0}	5.63×10^{0}	3.49×10^{0}	2.06×10^{0}	1.35×10^{0}
4.0×10^{0}	5.63×10^{0}	3.51×10^{0}	2.13×10^{0}	1.50×10^{0}
8.0×10^{0}	5.64×10^{0}	3.56×10^{0}	2.31×10^{0}	1.85×10^{0}
1.4×10^{1}	5.65×10^{0}	3.63×10^{0}	2.55×10^{0}	2.23×10^{0}
2.4×10^{1}	5.67×10^{0}	3.74×10^{0}	2.86×10^{0}	2.68×10^{0}
4.0×10^{1}	5.70×10^{0}	3.90×10^{0}	3.24×10^{0}	3.15×10^{0}
8.0×10^{1}	5.76×10^{0}	4.22×10^{0}	3.85×10^{0}	3.82×10^{0}
1.4×10^{2}	5.85×10^{0}	4.58×10^{0}	4.38×10^{0}	4.37×10^{0}

[a] Values obtained by setting sigma = 10^{-9}.

Table E.6. Continued

	Beta			
$1/u_A$	**1.0×10^{-3}**	**1.0×10^{-2}**	**6.0×10^{-2}**	**2.0×10^{-1}**
2.4×10^{2}	5.99×10^{0}	5.00×10^{0}	4.91×10^{0}	4.91×10^{0}
4.0×10^{2}	6.16×10^{0}	5.46×10^{0}	5.42×10^{0}	5.42×10^{0}
8.0×10^{2}	6.47×10^{0}	6.11×10^{0}	6.11×10^{0}	6.11×10^{0}
1.4×10^{3}	6.67×10^{0}	6.67×10^{0}	6.67×10^{0}	6.67×10^{0}
2.4×10^{3}	7.21×10^{0}	7.21×10^{0}	7.21×10^{0}	7.21×10^{0}
4.0×10^{3}	7.72×10^{0}	7.72×10^{0}	7.72×10^{0}	7.72×10^{0}
8.0×10^{3}	8.41×10^{0}	8.41×10^{0}	8.41×10^{0}	8.41×10^{0}
1.4×10^{4}	8.97×10^{0}	8.97×10^{0}	8.97×10^{0}	8.97×10^{0}
2.4×10^{4}	9.51×10^{0}	9.51×10^{0}	9.51×10^{0}	9.51×10^{0}
4.0×10^{4}	1.94×10^{1}	1.94×10^{1}	1.94×10^{1}	1.94×10^{1}
		Beta		
$1/u_B$	**6.0×10^{-1}**	**1.0×10^{0}**	**2.0×10^{0}**	**4.0×10^{0}**
4.0×10^{-4}	5.08×10^{-1}	3.18×10^{-1}	1.42×10^{-1}	4.79×10^{-2}
8.0×10^{-4}	5.08×10^{-1}	3.18×10^{-1}	1.42×10^{-1}	4.80×10^{-2}
1.4×10^{-3}	5.08×10^{-1}	3.18×10^{-1}	1.42×10^{-1}	4.81×10^{-2}
2.4×10^{-3}	5.08×10^{-1}	3.18×10^{-1}	1.42×10^{-1}	4.84×10^{-2}
4.0×10^{-3}	5.08×10^{-1}	3.18×10^{-1}	1.42×10^{-1}	4.88×10^{-2}
8.0×10^{-3}	5.09×10^{-1}	3.19×10^{-1}	1.43×10^{-1}	4.96×10^{-2}
1.4×10^{-2}	5.10×10^{-1}	3.21×10^{-1}	1.45×10^{-1}	5.09×10^{-2}
2.4×10^{-2}	5.12×10^{-1}	3.23×10^{-1}	1.47×10^{-1}	5.32×10^{-2}
4.0×10^{-2}	5.16×10^{-1}	3.27×10^{-1}	1.52×10^{-1}	5.68×10^{-2}
8.0×10^{-2}	5.24×10^{-1}	3.37×10^{-1}	1.62×10^{-1}	6.61×10^{-2}
1.4×10^{-1}	5.37×10^{-1}	3.50×10^{-1}	1.78×10^{-1}	8.06×10^{-2}
2.4×10^{-1}	5.57×10^{-1}	3.74×10^{-1}	2.05×10^{-1}	1.06×10^{-1}
4.0×10^{-1}	5.89×10^{-1}	4.12×10^{-1}	2.48×10^{-1}	1.49×10^{-1}
8.0×10^{-1}	6.67×10^{-1}	5.06×10^{-1}	3.57×10^{-1}	2.66×10^{-1}
1.4×10^{0}	7.80×10^{-1}	6.42×10^{-1}	5.17×10^{-1}	4.45×10^{-1}
2.4×10^{0}	9.54×10^{-1}	8.50×10^{-1}	7.63×10^{-1}	7.18×10^{-1}
4.0×10^{0}	1.20×10^{0}	1.13×10^{0}	1.08×10^{0}	1.06×10^{0}
8.0×10^{0}	1.68×10^{0}	1.65×10^{0}	1.63×10^{0}	1.63×10^{0}
1.4×10^{1}	2.15×10^{0}	2.14×10^{0}	2.14×10^{0}	2.14×10^{0}
2.4×10^{1}	2.65×10^{0}	2.65×10^{0}	2.64×10^{0}	2.64×10^{0}
4.0×10^{1}	3.14×10^{0}	3.14×10^{0}	3.14×10^{0}	3.14×10^{0}
8.0×10^{1}	3.82×10^{0}	3.82×10^{0}	3.82×10^{0}	3.82×10^{0}
1.4×10^{2}	4.37×10^{0}	4.37×10^{0}	4.37×10^{0}	4.37×10^{0}
2.4×10^{2}	4.91×10^{0}	4.91×10^{0}	4.91×10^{0}	4.91×10^{0}
4.0×10^{2}	5.42×10^{0}	5.42×10^{0}	5.42×10^{0}	5.42×10^{0}
8.0×10^{2}	6.11×10^{0}	6.11×10^{0}	6.11×10^{0}	6.11×10^{0}
1.4×10^{3}	6.67×10^{0}	6.67×10^{0}	6.67×10^{0}	6.67×10^{0}
2.4×10^{3}	7.21×10^{0}	7.21×10^{0}	7.21×10^{0}	7.21×10^{0}
4.0×10^{3}	7.72×10^{0}	7.72×10^{0}	7.72×10^{0}	7.72×10^{0}
8.0×10^{3}	8.41×10^{0}	8.41×10^{0}	8.41×10^{0}	8.41×10^{0}
1.4×10^{4}	8.97×10^{0}	8.97×10^{0}	8.97×10^{0}	8.97×10^{0}
2.4×10^{4}	9.51×10^{0}	9.51×10^{0}	9.51×10^{0}	9.51×10^{0}
4.0×10^{4}	1.94×10^{1}	1.94×10^{1}	1.94×10^{1}	1.94×10^{1}

Table E.7. Values of Water Dynamic Viscosity (after Matthess, 1982, p. 14)

Water Temperature (°F)	Dynamic Viscosity (10^{-3} P-sec)
32.0	1.7921
33.8	1.7313
35.6	1.6728
37.4	1.6191
39.2	1.5674
41.0	1.5188
42.8	1.4728
44.6	1.4284
46.4	1.3860
48.2	1.3462
50.0	1.3077
51.8	1.2713
53.6	1.2363
55.4	1.2028
57.2	1.1709
59.0	1.1404
60.8	1.1111
62.4	1.0828
64.4	1.0559
66.2	1.0299
68.0	1.0050
77.0	0.8937
86.0	0.8007
104.0	0.6560

Appendix F
Type Curves

Type curves for the five well functions in Appendix E, plotted on logarithmic paper, are presented in this appendix. A blank sheet of logarithmic paper of the same scale as that used in developing the type curves also is presented. The type curves may be enlarged and copied on overhead transparencies, and the sheet of logarithmic paper may be enlarged and copied for use by the reader in matching type curves and time-drawdown or distance-drawdown curves. Well function values in Appendix E may be plotted on logarithmic paper with scales selected by the user.

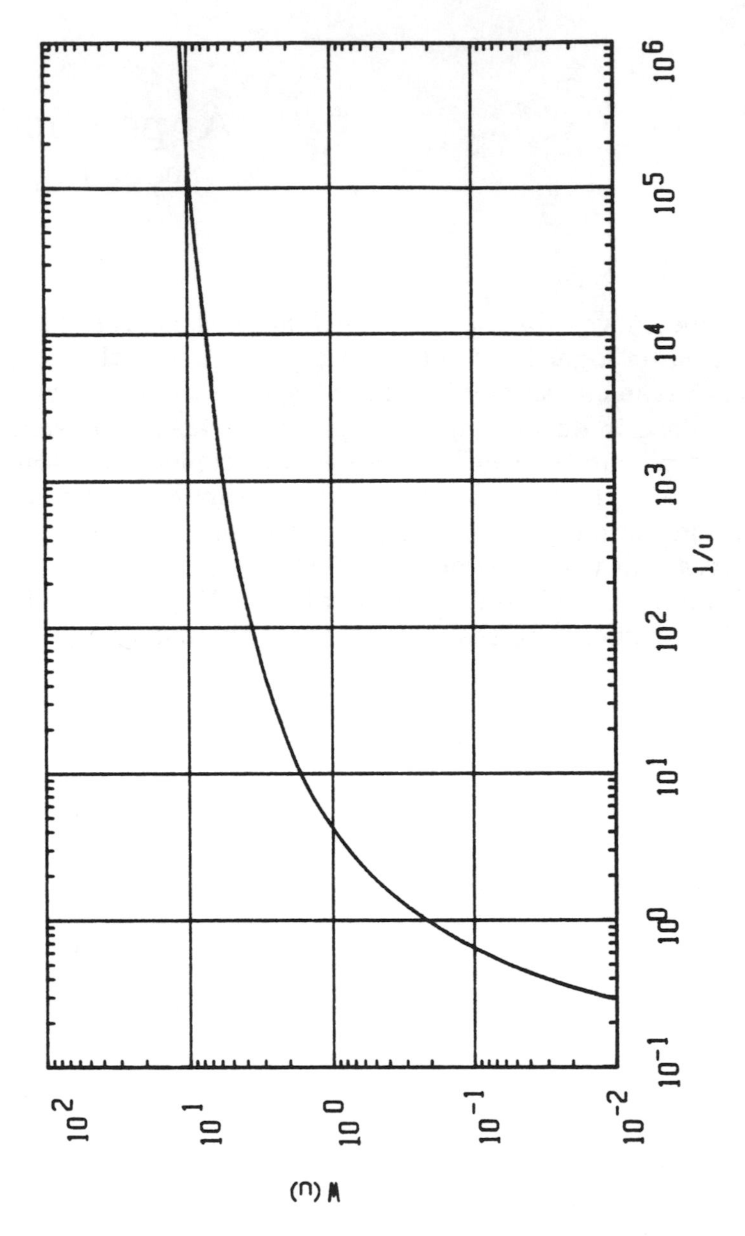

Figure F.1. W(u) versus 1/u type curve.

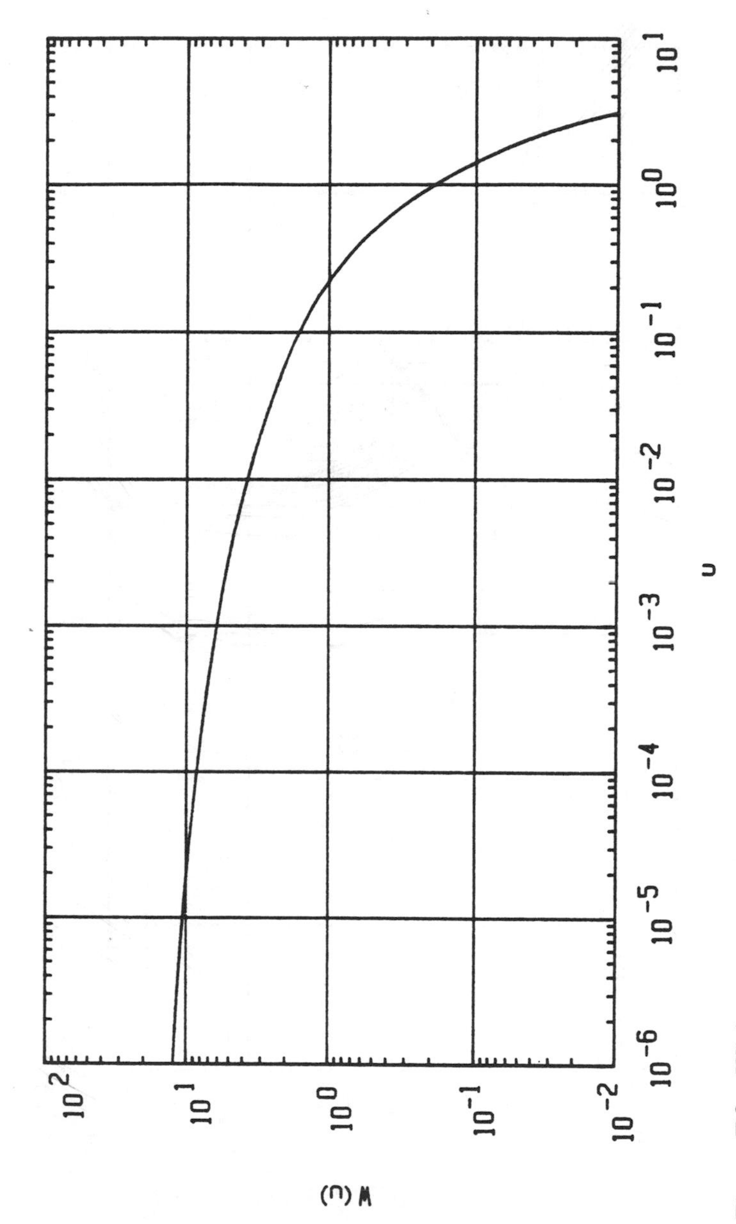

Figure F.2. W(u) versus u type curve.

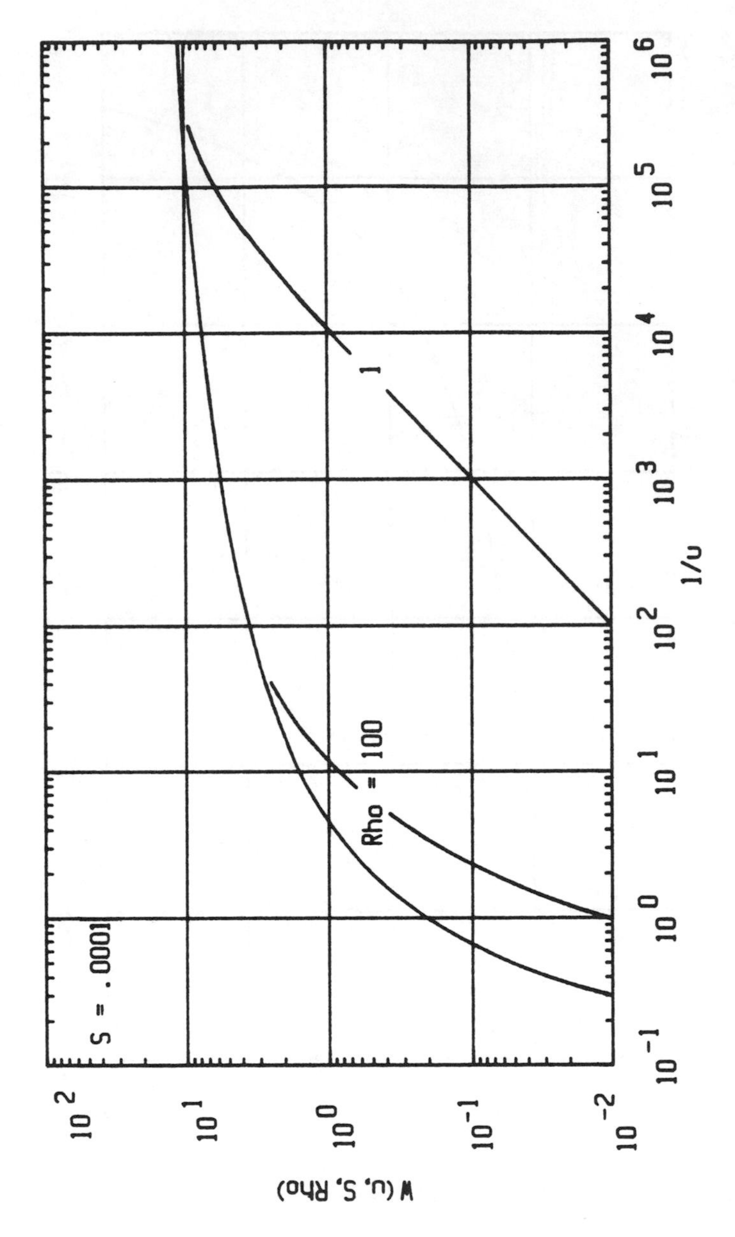

Figure F.3. W(u,S,Rho) versus 1/u family of type curves.

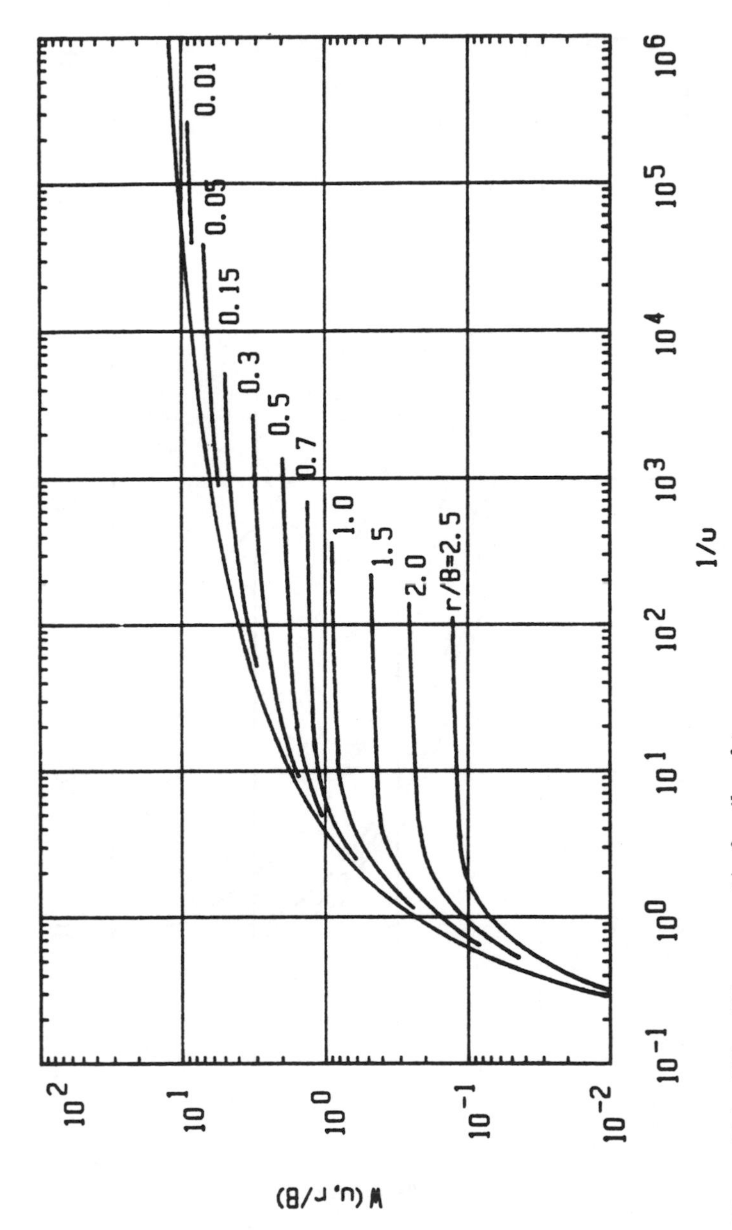

Figure F.4. W(u,r/B) versus 1/u family of type curves.

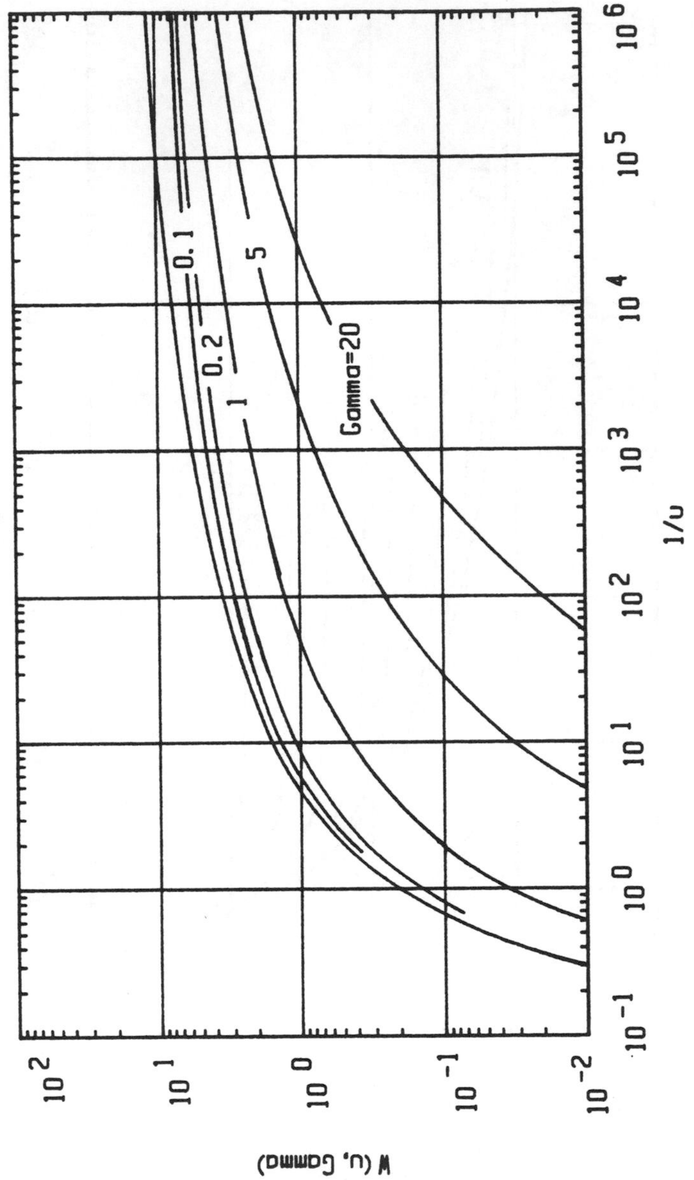

Figure F.5. W(u,Gamma) versus 1/u family of type curves.

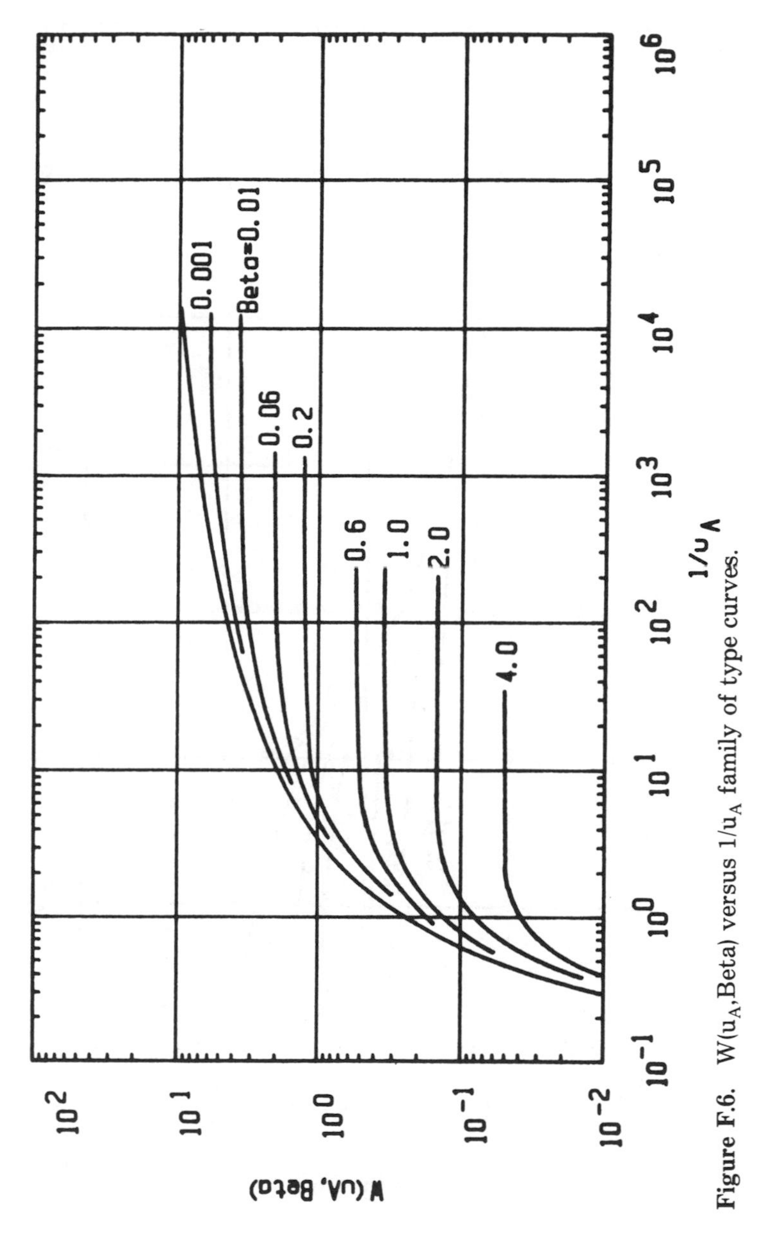

Figure F.6. $W(u_A, \text{Beta})$ versus $1/u_A$ family of type curves.

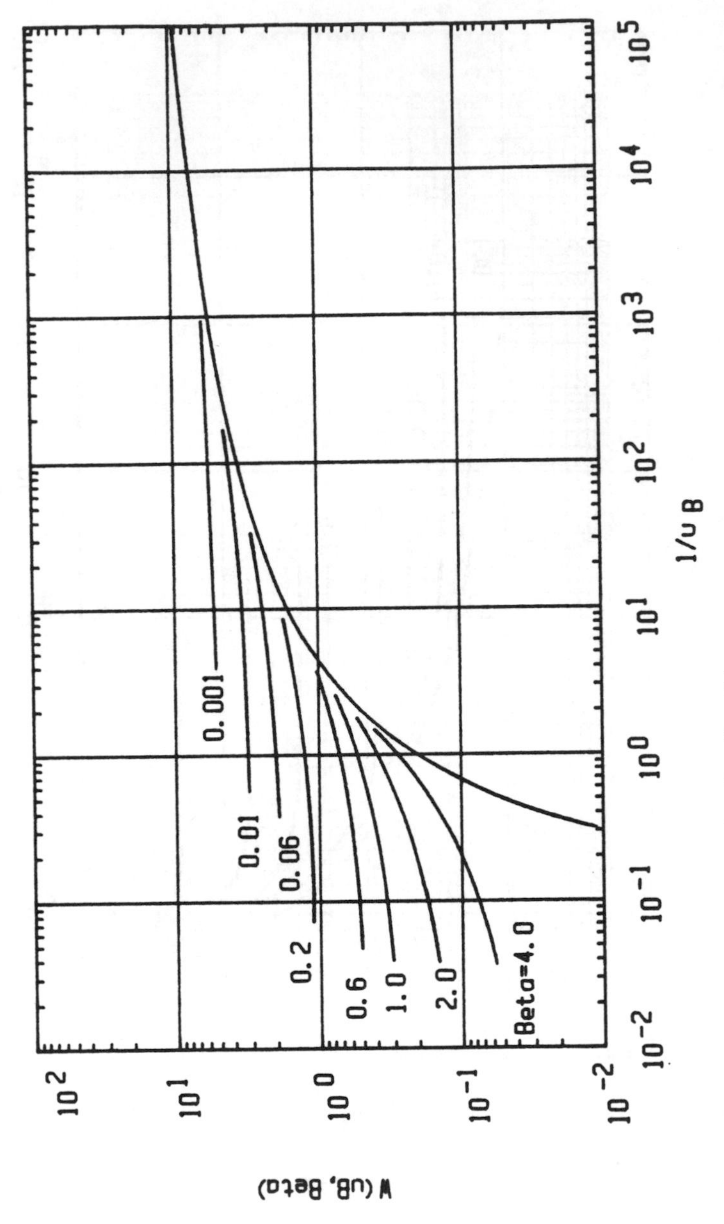

Figure F.7. $W(u_B,\text{Beta})$ versus $1/u_B$ family of type curves.

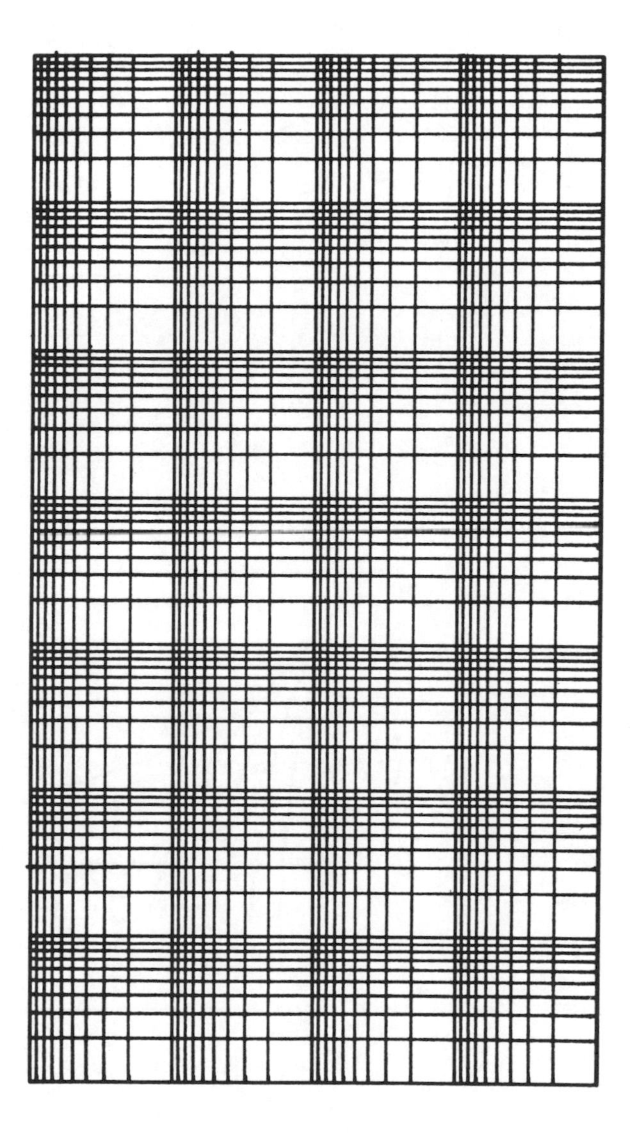

Figure F.8. Pumping test data graph paper.

References

Bentall, R., Comp. 1963. Shortcuts and Special Problems in Aquifer Tests. U.S. Geological Survey. Water-Supply Paper 1545-C.

Bredehoeft, J. D., and I. S. Papadopulos. 1980. A Method for Determining the Hydraulic Properties of Tight Formations. Water Resources Research. Vol. 16, No. 1.

Cooper, H. H., J. D. Bredehoeft, and I. S. Papadopulos. 1967. Response of a Finite-Diameter Well to an Instantaneous Charge of Water. Water Resources Research. Vol. 3, No. 1.

Cooper, H. H., and C. E. Jacob. 1946. A Generalized Graphic Method for Evaluating Formation Constants and Summarizing Well-Field History. Transactions of American Geophysical Union. Vol. 27, No. 4.

Davis, P. R., and W. C. Walton. 1982. Factors Involved in Evaluating Ground Water Impacts of Deep Coal Mine Drainage. American Water Resources Association. Water Research Bulletin. Vol. 18, No. 5.

Driscoll, F. G. 1986. Groundwater and Wells. Second Edition. Johnson Division. 1089 pp.

Ericson, R., and A. Moskol. 1986. Mastering Reflex. SYBEX, Inc. 336 pp.

Ferris, J. G., D. B. Knowles, R. H. Brown, and R. W. Stallman. 1962. Theory of Aquifer Tests. U.S. Geological Survey. Water-Supply Paper 1536-E. 174 pp.

Hantush, M. S. 1960. Modification of the Theory of Leaky Aquifers. Journal of Geophysical Research. Vol. 65, No. 11.

Hantush, M. S. 1964. Hydraulics of Wells. In Chow, Ven Te, Ed., Advances in Hydroscience. Vol. 1. Academic Press.

Hantush, M. S. 1966. Analysis of Data from Pumping Tests in Anisotropic Aquifers. Journal of Geophysical Research. Vol. 71, No. 2.

Hantush, M. S., and C. E. Jacob. 1955. Steady Three-Dimensional Flow to a Well in a Two Layered Aquifer. Transactions of American Geophysical Union. Vol. 36, No. 2.

Harris, J. M., and M. L. Scofield. 1983. IBM PC Conversion Handbook of BASIC. Prentice-Hall, Inc. 165 pp.

Hunt, B. 1983. Mathematical Analysis of Groundwater Resources. Butterworth & Co. 271 pp.

Jacob, C. E. 1944. Notes on Determining Permeability by Pumping Tests Under Water Table Conditions. U.S. Geological Survey. Mimeo. Report.

Jacob, C. E. 1946. Drawdown Test to Determine Effective Radius of Artesian Well. Proc. Am. Soc. Civil Engrs. Vol. 79, No. 5.

Jenkins, C. T. 1968. Computation of Rate and Volume of Stream Depletion by Wells. U.S. Geological Survey. Techniques of Water-Resources Investigations. Book 4, Chapter D1.

Kozeny, J. 1933. Theorie und Berechnung der Brunnen. Wasserkraft und Wassenwirtschaft. Vol. 28.

Kruseman, G. P., and N. A. De Ridder. 1976. Analysis and Evaluation of Pumping Test Data. International Institute for Land Reclamation and Improvement. Wageningen, The Netherlands. 200 pp.

Lohman, S. W. 1972. Ground-Water Hydraulics. U.S. Geological Survey. Professional Paper 708. 70 pp.

Matthess, Georg. 1982. The Properties of Groundwater. John Wiley & Sons, Inc. 406 pp.

Muskat, M. 1937. The Flow of Homogeneous Fluids Through Porous Medium. McGraw-Hill. 763 pp.

Neuman, S. P. 1974. Effect of Partial Penetration on Flow in Unconfined Aquifers Considering Delayed Gravity Response. Water Resources Research. Vol. 10, No. 2.

Neuman, S. P. 1975. Analysis of Pumping Test Data from Anisotropic Unconfined Aquifers Considering Delayed Gravity Response. Water Resources Research. Vol. 11, No. 2. *See* A Computer Program to Calculate Drawdown in an Anisotropic Unconfined Aquifer with a Partially Penetrating Well. Unpublished Manuscript. Dept. of Hydrology and Water Resources, Univ. of Arizona.

Neuman, S. P., and P. A. Witherspoon. 1969. Applicability of Current Theories of Flow in Leaky Aquifers. Water Resources Research. Vol. 5, No. 4.

Neuman, S. P., and P. A. Witherspoon. 1972. Field Determination of the Hydraulic Properties of Leaky Multiple Aquifer Systems. Water Resources Research. Vol. 8, No. 5.

Papadopulos, I. S. 1967. Drawdown Distribution Around a Large-Diameter Well. National Symposium on Ground-Water Hydrology, San Francisco, Calif. Proceedings.

Poole, L., M. Borchers, and K. Koessel. 1981. Some Common BASIC Programs. OSBORNE/McGraw-Hill. 193 pp.

Prickett, T. A. 1965. Type-Curve Solution to Aquifer Tests Under Water-Table Conditions. Ground Water. Vol. 3, No. 3.

Rathod, K. S., and K. R. Rushton. 1984. Numerical Method of Pumping Test Analysis Using Microcomputers. Ground Water. Vol. 22, No. 5.

Reed, J. E. 1980. Type Curves for Selected Problems of Flow to Wells in Confined Aquifers. U.S. Geological Survey. Techniques of Water-Resources Investigations. Book 3, Chapter B3.

Rorabaugh, M. I. 1956. Ground Water in Northeastern Louisville and Kentucky with Reference to Induced Infiltration. U.S. Geological Survey Water-Supply Paper 1360-B.

Rushton, K. R., and S. C. Redshaw. 1979. Seepage and Groundwater Flow. John Wiley & Sons, Ltd. 339 pp.

Sandberg, R., R. B. Scheiback, D. Koch, and T. A. Prickett.

1981. Selected Hand-Held Calculator Codes for the Evaluation of the Probable Cumulative Hydrologic Impacts of Mining. U.S. Dept. of Interior, Office of Surface Mining. H-D3004/030-81-1029F.

Stallman, R. W. 1963. Type Curves for the Solution of Single-Boundary Problems. In Bentall, R., Comp. Shortcuts and Special Problems in Aquifer Tests. U.S. Geological Survey. Water-Supply Paper 1545-C.

Stallman, R. W. 1971. Aquifer-Test Design, Observation and Data Analysis. U.S. Geological Survey. Techniques of Water-Resources Investigations. Book 3, Chapter B1.

Streltsova-Adams, T. D. 1978. Well Hydraulics in Heterogeneous Aquifer Formations. In Advances in Hydroscience. Ven Te Chow, Ed. Vol. 11. Academic Press.

Theis, C. V. 1935. The Relation Between the Lowering of the Piezometric Surface and the Rate and Duration of Discharge of a Well Using Ground-Water Storage. Transactions American Geophysical Union Vol. 16.

Walton, W. C. 1962. Selected Analytical Methods for Well and Aquifer Evaluation. Illinois State Water Survey. Bulletin 49.

Walton, W. C. 1963. Estimating the Infiltration Rate of a Streambed by Aquifer-Test Analysis. International Association of Scientific Hydrology. General Assembly, Berkeley.

Walton, W. C. 1970. Groundwater Resource Evaluation. McGraw-Hill. 664 pp.

Walton, W. C. 1985. Practical Aspects of Ground Water Modeling. National Water Well Association. Second Edition. 587 pp.

Wenzel, L. K. 1942. Methods for Determining Permeability of Water-Bearing Materials with Special Reference to Discharging-Well Methods. U.S. Geological Survey. Water-Supply Paper 887. 192 pp.

Witherspoon, P. A., and S. P. Neuman. 1972. Hydrodynamics of Fluid Injection. Underground Waste Management and Environmental Implications. Memoir No. 18. Am. Assoc. of Petroleum Geologists.

Index